P9-BJT-635

SYNTHETIC FOOD

OTHER BOOKS BY MAGNUS PYKE

Nothing Like Science
Slaves Unaware?
The Science Myth
Food Science and Technology
The Science Century
Food and Society
The Manual of Nutrition (HMSO)
Industrial Nutrition (MACDONALD & EVANS)
Nutrition (E.U.P.)
The Boundaries of Science (HARRAP)
Man and Food (WEIDENFELD)

SYNTHETIC FOOD

Magnus Pyke

PhD FIBiol FRIC FRSE

JOHN MURRAY

FIFTY ALBEMARLE STREET LONDON

Printed in Great Britain by
Spottiswoode, Ballantyne & Co Ltd
London and Colchester

0 7195 2060 6

CONTENTS

PREFACE

Among the varied commodities required to maintain civilised life, food has customarily been given special importance. It is perhaps arguable whether this paramountcy is justified. The complex life of a modern community would become equally impossible without adequate provision of fuel, housing, metals, paper, clothing, transport or a variety of chemical products. Food, while being of peculiar biological significance, is only one among many economic commodities. Technological effort has from earliest times been applied to improving the supply and quality of food just as it has to other forms of wealth. The application of lime to the soil to increase crop yields was introduced in remote antiquity, and ever since the advent of modern science, it has been recognised that chemistry, the science concerned with the composition of matter, can be applied to food as appropriately as it can to any other commodity. The situation has now been reached when a number of food components, including several of the vitamins and amino-acids, are being manufactured and marketed on a substantial scale.

It is clear from the information set out in the successive chapters of this book that present knowledge already enables the manufacture of food by chemical synthesis. This is interesting in itself. It is equally interesting to consider whether food can be synthesised sufficiently cheaply to allow it to compete with foods produced by the traditional biological means of agriculture and fishing or by more recently developed technological processes that are used for the manufacture of margarine, the preparation of leaf protein and fish flour or the propagation of unicellular organisms on hydrocarbons. But whether or not complete food commodities are ever prepared in practice entirely by chemical synthesis, there is little doubt that more and more of those parts and components of food, of which vitamins and amino acids are the forerunners, will be prepared in chemical factories. And even if the synthesis of the major food components cannot be carried out at a cost capable of competing with that of biological material, who can foresee what economic changes may occur in the future?

Besides contributing to the positive nutritional needs of the world's population in the future, synthetic food and synthetic flavour may also contribute to people's negative requirements. Eating is a pleasure as well as a biological necessity. For the motorised and automated citizens of tomorrow's world, for whom obesity is the most insidious form of malnutrition, it could become a diminishing pleasure. Here synthetic food could add to the joys of life, since, as I have described in Chapter 7, it is well within the bounds of practicability to synthesise agreeable and attractive food-stuffs guaranteed to do to those who eat them no good at all.

While the chemical formulae and references included in the following pages may be of use to research scientists and technologists, it is hoped that the narrative unfolded in this book will be of equal interest to those concerned with the implications of science on the commerce, the health and the welfare of the community.

M.P.

Cambus
Clackmannanshire
Scotland
November 1970

CHAPTER 1

SYNTHETIC PROCESSES EXIST

In all the numerous discussions of different methods of producing food, the possibility of manufacturing it by direct chemical synthesis is commonly overlooked. Yet the scientific knowledge of how this can be done is already available. Men have already subsisted in full health for 19 weeks on a mixture of pure chemical nutrients. It is true that the second part of the problem, that is, how to present these nutrients in the form of recognisable food-stuffs that would be accepted as an agreeable diet by normal consumers, still remains to be achieved, but if science allows us to make the ingredients it is surely not unreasonable to expect that technology will be able to fabricate them into attractive commodities. After all, the motive for doing so is a strong one.

The chemical diet fed for the 19 weeks to 24 carefully selected men[1]—actually, 24 men started the trial but only 15 finished; those who went back to 'real' food during the progress of the experiment, however, did so for other than medical reasons—is shown in Table 1.

This mixture of pure chemicals was prepared in the form of a liquid syrup. The men who submitted themselves to the test were volunteers, although it was thought prudent to lock them up in their quarters so that they should not be led into the temptation of eating natural food if they found themselves in the position of being able to obtain some. The purpose of the trial was to try to develop a diet which could be used under the very special and confined conditions of space travel. Finally, because of the particular purpose of the experiment, no consideration was paid either to the palatability of the mixture or to its cost. Yet in spite of these objections, the experiment showed that, if the occasion for doing so arises, a man can live on an entirely synthetic chemical diet. And, as I shall show in the following chapters, the knowledge of how to synthesise the nutrients of which foods are composed is available.

[1] Winitz, M. (1965), Graff, J., Gallagher, N., Narkin, A., and Seedman, D. A., *Nature, Lond.*, 205, 741.

Table 1

Composition of an artificial diet entirely composed of purified chemical compounds

Amino-acids			
l-Lysine·HCl	3·58 g	Sodium L-aspartate	6·40 g
l-Leucine	3·83 g	*l*-Threonine	2·42 g
l-Isoleucine	2·42 g	*l*-Proline	10·33 g
l-Valine	2·67 g	Glycine	1·67 g
l-Phenylalanine	1·75 g	*l*-Serine	5·33 g
l-Arginine·HCl	2·58 g	*l*-Tyrosine ethyl ester·HCl	6·83 g
l-Histidine·HCl·H_2O	1·58 g	*l*-Tryptophan	0·75 g
l-Methionine	1·75 g	*l*-Glutamine	9·07 g
l-Alanine	2·58 g	*l*-Cysteine ethyl ester·HCl	0·92 g
Water-soluble vitamins			
Thiamine·HCl	1·00 mg	*d*-Biotin	0·83 mg
Riboflavin	1·50 mg	Folic acid	1·67 mg
Pyridoxine·HCl	1·67 mg	Ascorbic acid	62·50 mg
Niacinamide	10·00 mg	Cyanocobalamin	1·67 μg
Inositol	0·83 mg	*p*-Aminobenzoic acid	416·56 mg
d-Calcium pantothenate	8·33 mg	Choline bitartrate	231·25 mg
Salts			
Potassium iodide	0·25 mg	Potassium hydroxide	3·97 g
Manganous acetate·$4H_2O$	18·30 mg	Magnesium oxide	0·38 g
Zinc benzoate	2·82 mg	Sodium chloride	4·77 g
Cupric acetate·H_2O	2·50 mg	Ferrous gluconate	0·83 g
Cobaltous acetate·$4H_2O$	1·67 mg	Calcium chloride·$2H_2O$	2·44 g
Sodium glycerophosphate	5·23 g	Sodium benzoate	1·00 g
Ammonium molybdate·$4H_2O$	0·42mg		
Carbohydrates			
Glucose	555·0 g	Glucono-δ-lactone	17·2 g
*Fats and fat-soluble vitamins**			
Ethyl linoleate	2·0 g	α-Tocopherol acetate	57·29 mg
Vitamin A acetate	3·64 mg	Menadione	4·58 mg
Vitamin D	0·057 mg		

* Provided daily in separate supplement.

The ability to synthesise foods from non-edible raw materials is clearly a matter of very great interest. Perhaps of equal significance, however, are the economic considerations. So far, the human race has obtained its sustenance from the plants and animals whose substance is derived from the biochemical mechanisms of nature. Whether or not the intellectual achievement by which the same components can now be produced by synthetic chemistry is to be of practical value depends on the economics of the various processes involved. There are several precedents showing that chemical synthesis can be a more economic method of producing a natural product than dependence on agriculture or animal husbandry. For example, in the last century, the discovery of how to synthesise indigo in the laboratory soon led to the manufacture not only of indigo, but of most other vegetable dyes as well, and to the ruin of the indigo farmers of Bengal. In the present century, hormones and numerous other drugs, at one time only to be derived from natural sources, are being economically produced by chemical synthesis. On the other hand, although the chemical manufacture of textile fibres has been one of the major scientific successes of the age, such fibres only constitute a proportion of the market and cotton and wool, produced as they always have been by the traditional operations of biology, continue to hold their own. It can, therefore, be expected that foods produced by chemical synthesis will at first take their place as a part of food production where they can be shown to possess some special economic or technical advantage.

Partial synthesis of food, as distinct from the use of synthetic components as part of an otherwise natural foodstuff, has of course already been attempted in several different ways. For example, inorganic ammonium salts have been converted into protein in several projects. During the 1939–45 war, surplus sugar, in the form of cane juice or molasses in the West Indies, was used as an energy source for yeast by which the ammonium salts could then be converted into protein. A special strain of yeast, *Torula utilis* var. *major*, capable of growing at a temperature of 37–38°C was used.[2] The sterilised molasses, sugar juice or solution of raw sugar was run continuously into a series of ten vessels, appropriate amounts of caustic soda, ammonium sulphate and ammonium superphosphate were added and the whole aerated through a simple arrangement of ceramic bubblers. The final liquor containing the yeast was continuously drawn off and the yeast cells separated on a centrifuge, washed and dried on heated rollers. The process was carefully planned[3] and, after several technical problems had been overcome, was eventually set into operation in Jamaica.

A similar process in which inorganic ammonium salts and wood,

[2] Colonial Food Yeast Ltd (1944), *Food Yeast*. HMSO, London.
[3] BWI Sugar Technologists (Oct. 1947), *Proc.* Sept., pp. 38.

both of which are inedible materials, are converted into protein, again through the intermediacy of growing yeast, has also been operated on a substantial commerical scale. During the manufacture of paper, very large volumes of so-called sulphite pulp liquor are produced. The disposal of this effluent is a major problem for the pulp mills, and strenuous efforts have been made to extract useful materials from it. Amongst its other components, the spent sulphite liquor from spruce wood contains from 15 to 22 g of sugar per litre,[4] of which about 80% is made up of hexose sugars and 20% of pentose sugars. The liquor must first be passed through a stripping still to reduce its content of sulphur dioxide to a concentration that the yeast will tolerate. It can then be used to feed yeast which is strongly aerated to achieve maximum growth. After this fermentation process, the yeast is separated and dried, usually by being passed over heated rollers. In spite of the fact that the main raw material, the spent sulphite liquor, is not only free but something which the paper makers are willing to pay to dispose of, it has not been found easy to produce a dried yeast sufficiently free from contamination to allow its use as an animal feed, and at a price to allow it to compete with other protein feeds of comparable composition. Nevertheless, the process undoubtedly represents the partial synthesis of a protein food, even though the synthetic operations are carried out biologically by yeast.

A third example of such mixed biological synthesis using components not normally susceptible to biological influences, and one which has attracted a good deal of attention, is the conversion of petroleum, together with ammonium salts and minerals, into protein, again by using the biosynthetic mechanisms of growing yeast.

There are two main processes for growing yeast on petroleum. On the one hand it can be propagated on a comparatively crude 'gas oil' fraction and the yeast subsequently freed from unmetabolised hydrocarbon residues by solvent extraction. Alternatively, highly purified n-alkanes[5] made up of C_{13} to C_{19} molecular chains can be prepared. When these are used as feedstock they are completely metabolised so that the recovered yeast when it is subsequently dried constitutes a protein feed-stuff uncontaminated by petroleum

[4] Inskip, G. C. (1951), Wiley, A. J., Holderby, J. M., and Hughes, L. P., *Ind. Eng. Chem.*, 43, 1702.

[5] Compounds with the structure $CH_3 \cdot CH_2 \cdot CH_2 \cdots CH_2 \cdot CH_3$.

residues. Both of these processes call for a strain of yeast capable of utilising hydrocarbons as its source of energy and growth. A research team made up of scientists from both Esso and the Swiss firm of Nestlé[6] screened more than 1000 species of bacteria and yeasts before finding one able to grow in satisfactory yield when fed on a substrate based on n-alkanes.

A great deal of development work has been done on the propagation of yeast by these means. Besides Esso and Nestlé, British Petroleum and their French subsidiary, BP France in Lavera, have worked on the problem, and the Institute of Gas Technology in Chicago is studying the fermentation of methane. The technique employed is basically similar to that used in the propagation of yeast on molasses or sulphite liquor. The exact conditions of temperature, aeration and feed rate have to be worked out, and where continuous fermentation is used the whole operation must be carried out under aseptic conditions similar to those used in the production of penicillin. The practical problem to be solved for all these processes of food manufacture, however, is that of producing a usable and nutritious material at a cost capable of competing with materials derived from less unusual and 'scientific' production methods.

The magnitude of the problem of economics which still remains to be dealt with even after all the technical difficulties have been overcome is shown in Table 2. Yet protein from petroleum, so-called 'single-cell protein' (SCP), is only partial synthesis. Although the components of the feedstock on which the yeasts or other micro-organisms are grown are unusual, the product is nevertheless derived from a biological process. This book is concerned with the direct synthesis of food components by chemical means. But while the technical problems in achieving these targets are interesting in their own right, their practical usefulness will also be affected by the cost of solving them.

In spite of the fact that, as shown in Table 2, the cost of protein in the form of yeast grown on petroleum is several times higher than the protein from peanut flour, soya bean or cottonseed, considerable scientific efforts are being devoted to the development of an economically viable process for making protein from petroleum by this method. In the next chapter, evidence is described suggesting that knowledge is now coming to light to enable protein to be directly synthesised, possibly from such readily available raw materials as methane gas

[6] Anon. (1967), *Chem. Eng. News*, 45 (2), 46.

Table 2

Cost of protein produced by growing micro-organisms on petroleum compared with costs of conventional protein sources

Product	*Protein content*	*Price per lb*	*Price per lb protein*
	%	US $	US $
Peanut flour	59	0.07	0.12
Soya bean	43	0.05	0.12
Cottonseed	50	0.05	0.10
Skim milk powder	36	0.15	0.41
Food yeast (*Torula*)	48	0.17	0.36
Protein from petroleum	—	—	0.35 (estimate)

and ammonia. There is still much work to be done before any claims can be made for a practical manufacturing process, but the importance of serious study of the feasibility of the principle involved is clear. Similarly, the synthesis of fat from petroleum or from coal is well established. Neither the quality of the fat produced nor its price were attractive when serious study ceased a generation ago. Yet in the generation past, advances in scientific understanding and technical ability have radically developed. Again, the evidence deployed in Chapter 3 indicates the possibility at least of there being an unworked vein of edible fat awaiting exploitation.

Of all food components, carbohydrates are the most plentiful. The scope for useful synthesis therefore seems least. Yet the history of science is full of neglected opportunities and the synthesis of 'formose', a mixture of sugars, was reported 80 years ago, with formaldehyde, readily available from coal or petroleum as a starting material. The neglect by industrial scientists of this discovery could well prove to be one more example of lack of imagination.

The vitamins were the last of the nutrients to be discovered, but they have been the first to be synthesised on a commerical scale. With all the other food components, there is current neglect of the opportunities offered. We have accumulated a substantial amount of scientific knowledge by which their synthesis could be achieved commercially. I hope that by the end of this short book I shall convincingly have demonstrated the case for a fundamental reappraisal of world policies towards synthetic food.

CHAPTER 2

SYNTHETIC PROTEIN

In the wealthy industrial societies the diet available to most people provides them with ample quantities of protein. Even among the less fortunate members of such communities the *total* supply of protein usually exceeds their minimum requirements; the composition of the protein as a whole, however, tends to change. It is generally found that as populations become richer the proportion of protein derived from animal foods, that is from meat, fish and dairy products, increases. Conversely, among the increasingly less fortunate people, the proportion of these more highly prized commodities diminishes, and the proportion of starchy foods of vegetable origin increases. In the West, these are commonly refined cereals, usually wheat, rye or maize; in the East, rice, and in Africa, manioc or cassava. This change in diet causes not only the total amount of protein to diminish but the composition of the protein in terms of its component amino-acids changes for the worse as well. As conditions become more and more stringent, not only does an absolute and relative deficiency of protein in terms of amino-acids occur, but there will finally be a shortage of calories as well, leading first to undernutrition and then, in its more prolonged and rigorous manifestation, to starvation.

The amounts of the different individual amino-acids required by the various groups of individuals—by infants, children, adolescents, adults, expectant and nursing mothers—are not necessarily present in one single food-stuff in the relative proportions required for the optimum nutrition of any one of these groups. The amounts needed are made up from the contribution made by different foods, each containing protein of differing composition, out of which a nourishing diet is commonly composed. It follows from this that chemical synthesis can make a contribution to protein nutrition in two ways. First, if those individual amino-acids known to be present in inadequate amounts in the cereal diets eaten by impoverished communities are synthesised separately, the supplementation of such diets merely by individual amino-acids will render the otherwise unsatisfactory protein already present adequate. Secondly, entire protein can be synthesised if the total protein content of the diet is inadequate.

It is only since the development of modern methods of chemical analysis, and the large-scale availability of synthetic amino-acids, that it has become possible to determine what the requirements of human beings—and of domestic animals and poultry, in whose nutrition scientists take an equal interest—actually are. Understanding of dietary requirements has been assimilated by a long series of laborious feeding trials in which mixtures of amino-acids were provided in place of protein, and the amount of one amino-acid after another reduced bit by bit until the limiting concentration of each was found. The first conclusion was that many of the amino-acids of which food proteins are composed are not essential to human or animal nutrition. Ten or so, however, are essential and the appropriate proportion of each must be provided by the diet for health to be maintained. These necessary dietary components are shown in Table 3.

Table 3

The amounts of different amino-acids required by different categories of people and by chickens, pigs and rats

Amino-acid	*Infants,*[1] *mg/kg*	*Women,*[2] *mg/day*	*Men,*[2] *mg/day*	*Young chicks,*[3] *% of diet*	*Young pigs,*[3] *% of diet*	*Young rats,*[3] *% of diet*
Histidine	34			0·3	0·3	0·4
Isoleucine	126	450	700	0·6	0·6	0·5
Leucine	150	620	1100	1·4	1·2	0·8
Lysine	103	500	800	1·0	1·1	1·0
Methionine	45	350	1100 (200*)	0·4	0·4	0·4
Phenylalanine	96	220	1100 (300†)	0·8	0·7	0·7
Threonine	87	305	500	0·6	0·6	0·5
Tryptophan	22	157	250	0·2	0·2	0·2
Valine	105	650	800	0·8	0·6	0·7
Arginine				1·2	0·3	0·2
Glycine				1·0		

* If ample cystine is available.

† If ample tyrosine is available.

[1] Holt, L. E., *et al.* (1960), *Protein and Amino-Acid Requirements in Early Life.* N.Y. Univ. Press.

[2] Nat. Acad. Sci. (1959), U.S. Nat. Res. Council Pub. 711. Washington.

[3] Almquist, H. J. (1969), *Protein and Amino-Acid Nutrition.* Academic Press, N.Y.

It will be noticed that the requirements of livestock, that is of growing pigs and chickens, quite apart from the needs of experimental rats, are expressed in terms of the percentage of each amino-acid in the diet as a whole. This arose from the fact that it was quickly discovered by animal nutritionists that the mixed feeding-stuffs used as animal diets gave more efficient growth—that is, the increase in the animals' weight per unit weight of feed consumed became greater—when comparatively small amounts of the amino-acids, methionine and lysine, were added. This is now readily explicable when we examine the figures shown in Table 4, in which the amino-acid composition of a number of common food commodities are listed.

Table 4

Amino-acid composition of the protein in a number of different food-stuffs[4]

(g per 16 g of N)

Amino-acid	*Wheat*	*Rice*	*Maize*	*Sorghum*	*Egg*	*Milk*	*Meat*
Histidine	2·1	2·2	2·5	1·8	2·4	2·6	3·4
Isoleucine	4·1	4·4	6·4	5·2	5·7	7·5	5·4
Leucine	6·8	8·2	15·0	13·2	8·8	11·0	8·1
Lysine	2·7	3·2	2·3	2·6	7·2	8·7	10·1
Methionine	2·0	1·8	3·1	0·5	3·8	3·2	2·6
Phenylalanine	5·0	4·6	5·0	4·6	5·7	4·4	4·4
Threonine	3·0	3·5	3·7	3·1	5·3	4·7	5·1
Tryptophan	1·3	0·1	0·6	1·0	1·3	1·5	1·1
Valine	4·3	6·3	5·3	4·7	8·8	7·0	7·0
Arginine	4·3	8·6	4·8	3·2	6·5	4·2	6·9

It is clear from the figures in Table 4 that people—or livestock—who eat a diet in which the protein, perforce, is mainly obtained from cereals can expect to go short of the amino-acid, lysine, since cereal protein only contains 2–3 g of lysine per 16 g of N, whereas the proportion of lysine in animal-protein foods ranges from 7 to 10 g per 16 g of N and the need for lysine is comparatively great. Further,

[4] Hopper, T. H. (1958), *Processed Plant Protein Foodstuffs*, p. 877. Academic Press, N.Y.

as has been shown in experiments with laboratory animals,[5] whereas the nutritional value of diets made up mainly of wheat is improved by adding lysine alone, those in which maize is the principal source of protein are improved by adding both lysine and tryptophan. Rice diets are also improved by supplementations of these amino-acids and, it has been found, by threonine as well.

Although there is no cause to doubt that this reasoning applies to man, almost all the experiments which have been done to examine the effect of adding synthetic amino-acids to predominantly cereal diets or diets comprising low levels of animal protein have been done on animals. Typical results[6] with rats show that whereas young rats fed on a diet largely made up of cooked rice grew 85 g in 5 weeks, the addition of 0·05% of *l*-lysine[7] allowed a parallel group of animals to grow 106 g, and the addition of 0·10% brought about growth of 118 g in 5 weeks. It should be said, however, that when the amount of lysine added was further increased to 0·20%, far from causing still further acceleration in growth, it was actually accompanied by a fall in growth to 93 g.

The same sort of results were obtained when diets mainly composed of corn meal (maize) and when wheat (in the form of white bread) were used. Maximum advantage occurred when 0·05% of L-lysine hydrochloride was added to the bread diet and when 0·025% was added to the maize diet. And again, further increase in the amounts of lysine added caused the rats to grow less. These results exemplify both the usefulness of supplementing diets lacking in animal foods with synthetic amino-acids, most especially lysine, and also one at least of the problems that arise when supplementation is undertaken. There is no doubt that the nutritional value of individual protein foods, notably vegetable foods and particularly maize, is improved by the addition of lysine and sometimes methionine and threonine as well. Yet when the protein content of the diet is low, the addition of a single amino-acid—and particularly addition in excess—may cause an imbalance in the total amino-acid content of the diet as a whole so that growth, instead of being improved, is actually impeded.[8]

[5] Howe, E. E. (1961), U.S. Nat. Acad. Sci., Pub. 843. Washington.

[6] Rosenberg, S. R. (1959), *Agric. and Food Sci.*, 7, 316.

[7] The difference between *l*-lysine, *d*-lysine and lysine hydrochloride is explained on p. 12.

[8] Gullino, P. (1955), Winitz, M., Boinbaum, S. M., Cornfield, J., Otey, M. G., and Greenstein, J. P., *Arch. Biochem. and Biophys.*, 58, 253.

Clearly, to benefit the nutritional well-being of an impoverished community by adding lysine, either with or without other amino-acids, is a procedure requiring some degree of skill. At least, however, the supplementation of food containing too little protein, or protein whose amino-acid composition is unbalanced, has not been accompanied by demonstrable harm. It is also encouraging to record that the US Food and Drugs Administration[9] has reached the conclusion that synthetic lysine at least can be described as GRAS, that is, 'generally recognised as safe'.

The usefulness of adding synthetic amino-acids to animal feeding-stuffs has already been demonstrated on the harsh touchstone of economic returns. Compounders of feed for pigs and poultry purchase a substantial tonnage, and even in the 1950s the production of synthetic methionine in the United States alone amounted to 500 tons a year.[10] The value of adding synthetic amino-acids to human diets cannot be so directly assessed. In Western countries where wheat is the main dietary cereal, a lack of lysine could be anticipated, although, since wheat protein is in many respects more satisfactory in its nutritional composition than the protein of either rice or maize and an absolute deficiency of protein is infrequent, clear evidence of a deficiency of lysine is not easy to demonstrate. There is a very much stronger argument for adding synthetic amino-acids to the diets of Eastern and African peoples who use rice or, worse still, starchy roots as their main source of calories.

But whether or not it can be economic to manufacture amino-acids, and particularly lysine, as a fine chemical will depend on several factors besides the scientific feasibility of carrying out the chemical synthesis. First, of course, is the relative cost of the manufactured product in comparison with the cost of natural lysine-containing protein from natural sources. Then again, chemical synthesis is not the only way of producing amino-acids. For example, lysine can be manufactured by a fermentation process[11] at a cost that is claimed to be as low as, or even lower than, that of synthetic products.

But a more subtle means of increasing the lysine content of the predominately cereal diets eaten by peoples in developing countries

[9] U.S. Federal Register, 20 Nov. 1959.

[10] Rosenberg, H. R. (1957), *Agric. Food Chem.*, 5, 697.

[11] U.S. Patents 2,771,396 and 2,841,532.

has been achieved by a group of American plant geneticists.[12] These workers have succeeded in breeding a strain of maize in which the protein contains substantially more lysine than is present in normal varieties. This represents yet another example in which the science of biology competes—and often competes successfully—with the science of chemistry to attain a desired result. Lysine, however, together with other nutritionally important amino-acids, has been manufactured, as I shall now describe, by chemical synthesis. Since the chemical manufacture of all such organic molecules is inevitably highly technical, the following accounts may seem somewhat obscure to readers unversed in organic chemistry. The implications of the inevitably intricate manufacturing methods on the even newer discoveries described later in the chapter are, however, of remarkable potential significance.

Since lysine is the amino-acid most likely to be lacking in poor diets, considerable attention has been given to methods of manufacturing it on a commercial scale.[13] Its chemical structure is shown below:

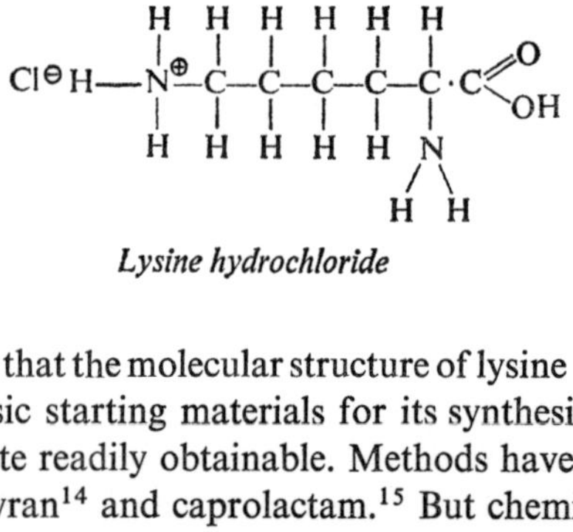

```
           H  H  H  H  H  H
           |  |  |  |  |  |      O
  Cl⊖ H—N⊕—C—C—C—C—C·C<
           |  |  |  |  |  |      OH
           H  H  H  H  H  N
                          / \
                         H   H
```

Lysine hydrochloride

It can be seen that the molecular structure of lysine is comparatively simple. The basic starting materials for its synthesis have also been found to be quite readily obtainable. Methods have been developed using dihydropyran[14] and caprolactam.[15] But chemical synthesis by either of these methods yields a mixture of two lysine molecules, one in which the structure is of the so-called *laevo* form, which is biologically active, and the other, identical in all respects except that the

[12] Mertz, E. T. (1964), Bates, L. S., and Nelson, O. E., *Science*, 145, 279.

[13] Anon. (1955), *J. Agric. Food Chem.*, 3, 647; Anon. (1956), *Chem. Week*, 79 (25), 52.

[14] Gaudry, R. (1948), *Can. J. Res.*, 26B, 387; U.S. Patents 2,498,300, 2,556,917, 1,564,647–9.

[15] Eck, J. C. (1934), and Marvel, C. S., *J. Biol. Chem.*, 106, 387.

molecule is 'folded' differently in the so-called *dextro* configuration, which is biologically unavailable. Further study had therefore to be undertaken to separate only the nutritionally valuable *laevo* form. This has been achieved[16] and synthetic *l*-lysine monohydrochloride can be produced to a degree of purity of 95% (the impurity being a residue of lysine monohydrochloride in the *d*-form).

The significance of the production of lysine, not as the free amino-acid but in the form of the monohydrochloride (the dihydrochloride will also occur if the conditions under which the lysine is crystallised are not adjusted appropriately), must be taken into account in assessing the nutritional potency of the synthesised material. Lysine is quite strongly basic in its chemical reaction and is difficult to prepare in the free form. It is important to remember, therefore, that when it is used, as it most commonly is, as a salt of hydrochloric acid, the hydrochloric acid moiety of lysine monohydrochloride makes up almost exactly 20% of the weight of the molecule. Consequently, even when the lysine is entirely of the nutritionally active *laevo* form only 80% of the amount added will be available.

The manufacture and marketing of synthetic lysine has already become a comparatively large-scale operation. According to the US Tariff Commission, in 1958, 121 000 lb of *l*-lysine hydrochloride was produced in the United States and sold at an average cost of $9.08 per lb. In 1959, the amount produced had more than doubled to 273 000 lb and the cost fallen to $5.13 per lb. In 1961, production exceeded 300 000 lb and the price averaged $4.33 per lb.

Methionine, which was the first amino-acid to be synthesised on a commercial scale, is also the only amino-acid of which the two optically active forms, the *laevo* and the *dextro* isomers, are equally well utilised in human and animal nutrition. The chemical structure of methionine is shown below:

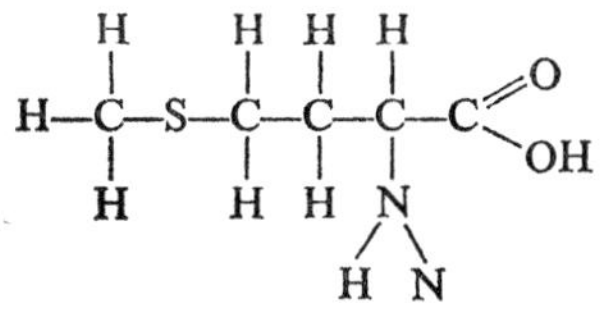

```
    H       H   H   H
    |       |   |   |      O
H—C—S—C—C—C—C<
    |       |   |   |      OH
    H       H   H   N
                   / \
                  H   N
```

Methionine

[16] U.S. Patents 2,556,907; 2,657,230; 2,859,244.

The synthesis of methionine can be carried out by the combination of acrolein with methyl mercaptan to produce an intermediary compound which is then converted into the aminonitrile or the hydantoin and thence into methionine.[17]

For more than a decade methionine has been manufactured on a commercial scale and used, not to provide human food directly, but as an enrichment of animal feeds to increase the efficiency by which meat is produced from cereal rations. Already in 1954, about 1 000 000 lb of synthetic methionine were manufactured in the United States at a cost $2.91 per lb, according to the figures of the US Tariff Commission. By 1960, the output had substantially increased and the price fallen to $1.35 per lb.[18]

Synthesis of the amino-acid, threonine, like that of lysine, again presents the organic chemist with the problem of separating the desirable isomer from the mixture of isomers which are produced simultaneously during synthesis. As happens so often in life, the unwanted form—*dl*-allothreonine—is much more readily produced than the usable *dl*-threonine. A method has, however, been developed by three Japanese chemists in which cupric glycinate, used as a starting material, is reacted with acetaldehyde.[19] The product contains about 25% of the unwanted *dl*-allothreonine from which the *dl*-threonine has to be separated. The chemical structure of threonine is shown below:

```
     H  H    H
     |  |    |    //O
  H—C—C——C—C
     |  |    |    \OH
     H  O    N
        |   / \
        H  H   H
```

Threonine

The projected cost of synthetic threonine of 75% purity manufactured on a large scale has been estimated at between $1.75 and $3.10 per lb.[20]

Of those amino-acids whose synthesis promises to be of immediate direct usefulness, tryptophan has so far proved to be the most expensive. This is because indole, which is required as the

[17] U.S. Patents 2,516,635;

[18] White, H. C. (1960), Conf. on Protein Needs, Washington, D.C.

[19] Sato, N., Okawa, K., and Akabori, S. (1959), *Bull. Chem. Soc., Japan,* 30, 937.

[20] Fox, S. W. (1963), *Food Tech.*, 17 (4), 22.

starting material for the most practicable method of synthesis so far worked out,[21] is itself expensive. An alternative synthetic route via phenylhydrazine, acrolein and diethylacetamidomalate, is also not without drawbacks.[22] Nevertheless, the estimated future cost of tryptophan, provided a large-scale market were available, has been put at $4.70 to $8.00 per lb.[23]

The knowledge and ingenuity of organic chemists is such that none of the 18 or so constituent amino-acids of protein could not be manufactured, given adequate demand and economic reward. It has been seriously estimated[24] that given sufficiently large demand any of the L-amino-acids could be synthesised for something between $1.00 and $4.00 per lb. Dr White, of the American pharmaceutical firm Dow Chemical Company[25] has estimated that both of the amino-acids *l*-isoleucine and *dl*-phenylalanine could be manufactured for $1.05 to $1.80 per lb.

The synthesis of individual amino-acids has thus already been achieved, and several, including lysine, methionine, threonine and tryptophan, are manufactured profitably on a large scale. But the enrichment of animal feeds (which increases the rate of pig and poultry meat production) as well as the enhancement of the nutritional value of cereal or starchy human diets (the source of 'second-class' protein) does not represent food synthesis in its fullest meaning. Further, the separate synthesis of individual amino-acids, one after another, demands intricate organic chemistry and their fabrication in bulk presents, as we have seen, quite a complex exercise in technological and economic logistics.

Yet if this is so for the separate amino-acids, which, while of diverse configuration one from another are essentially small and simple molecular structures, the possibility of artificially fabricating the immense complexity of the subtle polymers constituting protein might be thought to be entirely beyond the bounds of the possibilities of chemical synthesis. Some idea of the intricacy of the molecular configuration of protein can be obtained from Figure 1[26] in which

[21] Snyder, H. R. (1964), and Smith, C. W., *J. Am. Chem. Soc.*, 66, 350, 500.
[22] U.S. Patents 2,516,332; 2,523,746.
[23] Fox, S. W. (1963), *Food Tech.*, 17 (4), 22.
[24] Howe, E. E. (1961), U.S. Nat. Res. Coun., Pub. 843, 495.
[25] White, H. C. (1960), Conf. on Protein Needs, Washington, D.C.
[26] Thompson, E. O. P. (1957), *The Physics and Chemistry of Life*, p. 87. Bell, London.

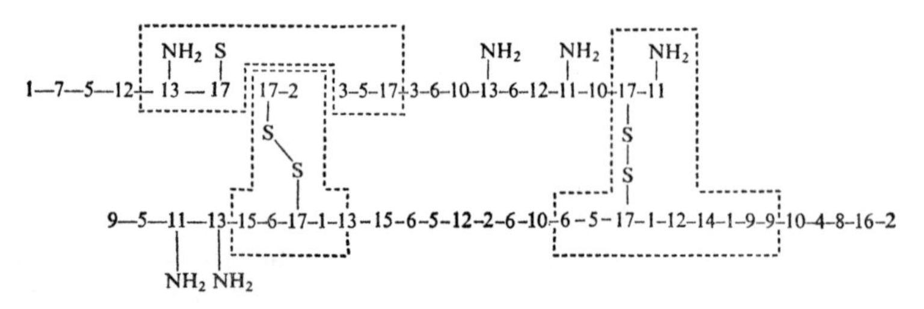

Figure 1 The insulin molecule. Complete molecule of insulin is depicted in the structural diagram above. Each amino-acid in the molecule is represented by a number and may be identified by reference to the table below. The molecule consists of 51 amino-acid units in two chains. One chain (top) has 21 amino-acid units; it is called the glycyl chain because it begins with glycine (1). The other chain (bottom) has 30 amino-acid units; it is called the phenylalanyl chain because it begins with phenylalanine (9). The chains are joined by sulphur atoms (S–S).

Amino-acid groups in insulin

Key number	*Name*	*Number of groups*	
		In phenylalanyl	*In glycyl*
1	Glycine	3	1
2	Alanine	2	1
3	Serine	1	2
4	Threonine	1	0
5	Valine	3	2
6	Leucine	4	2
7	Isoleucine	0	1
8	Proline	1	0
9	Phenylalanine	3	0
10	Tyrosine	2	2
11	Asparagine	1	2
12	Glutamic acid	2	2
13	Glutamine	1	2
14	Arginine	1	0
15	Histidine	2	0
16	Lysine	1	0
17	Cystine	2	4
		30	21

is shown the structure of the comparatively simple molecule of the protein, insulin, which is composed of only 771 atoms. For his brilliant work leading to the elucidation of this structure (not for the synthesis of the protein, which was entirely beyond the limits of the chemical knowledge of his day) Sanger was awarded a Nobel Prize in 1958.

It is appropriate at this point to quote from a paper written by Professor Sidney Fox of the University of Miami, Florida.[27] In discussing the possible way in which so complex a material as protein is synthesised in nature he wrote: 'We had been influenced by the chemists' typical assumption that, whereas simple compounds require simple synthesis, complex compounds, for example, proteins require complex processes.' This typical assumption would require that, first of all, two of the amino-acids known to be present in protein should be separately synthesised and then joined together by the appropriate atomic linkage. Next, a third amino-acid should be synthesised and attached to the other two, and so the process would continue until at last the entire molecule was built up—a daunting task for nature as it would be for man, no matter how expert his chemical virtuosity.

Fox and his colleagues had several reasons for abandoning chemical orthodoxy and adopting the scientific 'heresy' from which a new conception of the possibility of protein synthesis is arising. In nature the 'difficult' synthesis of complex proteins appeared to have occurred with equal complexity in the primitive forms of life which evolved early in evolutionary history as in much later and more complex forms down to the mammals and man himself. The amino-acid composition of proteins found in algae, bacteria, protozoa, invertebrates and mammals alike all contain the same 18 amino-acids, and, more surprising still, in relative proportions not differing very widely in the protein of the different groups of organisms.

In 1913, a German chemist, Löb,[28] observed that when an electric discharge was passed through a mixture of carbon monoxide gas, ammonia and water—all of which are deduced to have been present in the Earth's atmosphere before life began—glycine, the simplest of all the amino-acids, was formed. In 1955 Miller[29] found that in this type of experiment, not one, but several different amino-acids were produced simultaneously. This discovery attracted a great deal of interest because it seemed to show one way at least in which amino-acids, which are the basic components of protein, the essential

[27] Fox, S. W. (1965), *Nature, Lond.*, 205, 328.
[28] Löb, W. (1913), *Chem. Ber.*, 46, 690.
[29] Miller, S. L. (1955), *J. Am. Chem. Soc.*, 117, 528.

substance of living things, could have evolved from non-living chemical components. Pursuing this line of research in 1964, Harada and Fox[30] found that amino-acids could be produced from the kinds of gases present in the earth in pre-biological times not only by the action of electric discharges (such as lightning), but also by ultra-violet light, X-rays, nuclear radiation and heat. In one of their typical experiments, Harada and Fox passed methane gas (the explosive 'fire-damp' gas encountered in coal-mines) and ammonia through several different kinds of solids common on the crust of the Earth throughout the ages. For example, quartz sand was used, as well as silica gel, alumina and volcanic sand collected from Stromboli. These experiments were carried out at several carefully controlled temperatures between 950 and 1050°C. The remarkable results from these experiments can be seen from Table 5. Except for cystine, methionine, histidine and tryptophan, all the amino-acids common to protein were produced simultaneously in the thermal reaction over silica. Cystine and methionine contain sulphur, and since no sulphur-containing gas was used in the experiment their synthesis was not possible. Analysis for the final amino-acids, histidine and tryptophan, was not carried out.

But the simultaneous synthesis of the 18 amino-acids found in the proteins of living creatures extending right across the phylogenetic spectrum—and thus demonstrating the underlying unity of the life process throughout creation—is not the only remarkable observation brought into focus by Fox and his co-workers. In an experiment carried out in the summer of 1963 on the lava field of the volcano of Kapolo on the Island of Hawaii, Fox demonstrated that if an appropriate mixture of separate amino-acids was exposed for 3–4 h to a temperature of 170°C ,which was the temperature found in the lava, the separate amino-acids would link together to form a polymer similar in many respects to protein. The initial purpose of this research was to speculate on the origin of protein, and hence of life itself. Fox quickly realised as well that, by initiating in the laboratory what nature may have done in pre-biological ages long past, it might be possible to synthesise food from basic raw materials—methane, ammonia and simple sulphur compounds—and to do so without having to carry out complex synthesis of individual amino-acids or

[30] Harada, K. (1964), and Fox, S. W., *Nature, Lond.*, 201, 335.

Table 5

Amino-acids produced from simple gases by heat in the presence of silica and by electric discharge (%)

	Fox and Harada's synthesis by heat			*Miller's synthesis using electric discharge*	
Amino-acids	*Silica sand* 950°C	*Silica gel* 950°C	*Silica gel* 1050°C	*Spark discharge*	*Silent discharge*
Aspartic acid	3·4	2·5	15·2	0·3	0·1
Threonine	0·9	0·6	3·0		
Serine	2·0	1·9	10·0		
Glutamic acid	4·8	3·1	10·2	0·5	0·3
Proline	2·3	1·5	2·3		
Glycine	60·3	68·8	24·4	50·8	41·4
Alanine	18·0	16·9	20·2	27·4	4·7
Valine	2·3	1·2	2·1		
Alloisoleucine	0·3	0·3	1·4		
Isoleucine	1·1	0·7	2·5		
Leucine	2·4	1·5	4·6		
Tyrosine	0·8	0·4	2·0		
Phenylalanine	0·8	0·6	2·2		
α-Aminobutyric acid	0·6			4·0	0·6
β-Alanine	?	?	?	12·1	2·3
Sarcosine				4·0	44·6
N-methylalanine				0·8	6·5

to tackle the daunting task of constructing piece by piece the intricate molecular structure of a protein polymer.

The procedure for synthesising 'proteinoid' is remarkably simple. A mixture of dry amino-acids, in which the precise combination can be varied provided an adequate proportion of aspartic and glutamic acids is present, is heated to 170°C for 6 h and the subsequent pale-coloured polymers purified by the normal methods used for the separation of protein. 'Proteinoids' of molecular weights of 3500 up to 8500 have been obtained which are of the order of magnitude of some natural proteins, although smaller than those of most. The use of such biologically extreme conditions as a temperature of 170°C can be modified by adding mixtures of hypophosphite and polyphosphoric acid.[31]

[31] Hoogland, P. D., quoted by Fox, S. W. (1965), *Nature, Lond.*, 205, 328.

The discovery of 'proteinoids', synthesised so simply, first, by putting energy of one sort or another into mixtures of such chemical compounds, and then submitting the mixture of amino-acids formed to further heating under appropriate conditions, has opened up a new field of possibilities. Already, studies have been made to discover how 'proteinoids' thus produced form themselves into globules and cellular shapes which may bud and divide and exhibit many of the characteristics of living cells. 'Proteinoids' have also been found to be susceptible to breakdown by proteolytic enzymes which are characteristically taken only to be able to operate on protein. It would indeed be a new and potentially fruitful avenue for progress if food synthesis could be achieved by mimicking the basically simple evolutionary process by which life first began rather than by the technical complexities of the higher organic chemistry.

So far, protein has *not* been synthesised on a commercial scale. 'Proteinoids' have served as food for micro-organisms and have even been used as a component of the diet of experimental rats. But much has to be done before we can say whether protein substitute can be manufactured thus, or even whether 'proteinoid' is safe as a food component, quite apart from whether a 'proteinoid' possessing the composition and structure compatible with the high nutritional value of, say, egg can be prepared. There is a possibility (I would not put it higher) that an unnatural combination or imbalance of amino-acids in a "proteinoid" could actually be toxic.[32] One instance of this may occur if there is an excess of leucine in the diet; the adverse effect can then be put right by increasing the relative proportion of isoleucine.[33] High levels of some amino-acids, however, increase the dietary requirement for the vitamin, niacin. Several research workers have reported that it is not always possible to put right the unsatisfactory nutritional value of inadequate proteins (for example, burned fish meal, maize gluten meal, gelatin or blood meal[34]) by supplementing them with individual amino-acids. The toxic or retarding action of amino-acid imbalance, such as it is, is most acute when the total protein content of the diet is low, but it is under just such

[32] Harper, A. E. (1956), *Nutrit. Rev.*, 14, 225.

[33] Harper, A. E. (1955), Benton, D. A., and Elvehjem, C. A., *Arch. Biochem. Biophys.*, 57, 1.

[34] Clandiniu, D. R. (1949), *Poultry Sci.*, 28, 128; Grau, C. R. (1946), *J. Nutr.*, 32, 303; Hier, S. W. (1944), Graham, C. E., and Klein, D., *Proc. Soc. Exp. Biol. Med.*, 56, 187; Grau, C. R. (1944), and Almquist, H. J., *Poultry Sci.*, 23, 486.

circumstances that synthetic 'proteinoid' would be most likely to be used. Bressani *et al.*[35] have already observed that when they added the amino-acid, methionine, in what they had estimated to be the amount necessary to supplement the inadequate composition of the cereal protein upon which a group of malnourished infants were subsisting, far from there being an improvement in the growth and build-up of the tissues of the children, the amino-acid supplementation produced a positively detrimental effect.

Nevertheless, in spite of these biochemical considerations, many experiments on animals have shown that the nutritional quality of cereal protein can be improved by adding measured amounts of synthetic protein amino-acids. An index called the 'protein efficiency ratio'—PER—is most commonly used to measure the nutritional value of a particular protein. PER is defined as the increase in the weight of weanling rats fed for 28 days on a diet containing the protein under test divided by the weight of protein eaten.[36] Every experiment is compared with one in which the milk protein, casein, is used as a standard and the PER value for the casein adjusted to the arbitrary value of 2·5. Table 6 shows the effect of supplementing cereals with added synthetic lysine on their nutritional value as measured by the PER feeding test on baby rats.

Table 6

The effect of added synthetic lysine on the nutritional value of cereals as measured by the 'protein efficiency ratio' (PER) test[37]

Cereal	*PER of original grain*	*Amount of l-lysine hydrochloride added, %*	*PER of fortified grain*
Millet	0·7	0·30	2·1
Sorghum	0·7	0·20	2·2
Wheat	1·3	0·10	1·7
Maize	1·5	0·10	2·0
Rice	1·5	0·05	2·3
Barley	1·6	0·05	1·9

[35] Bressani, R. (1958), Wilson, D., Béhar, M., and Scrimshaw, N. S., *Fed. Proc.*, 17, 471.

[36] Chapman, D. G. (1959), Castillo, R., and Campbell, J. A., *Can. J. Biochem. Physiol.* 37 (5), 679.

[37] Howe, E. E. (1967), *Rep. Pres. Sci. Adv. Committee*, 2, 319.

Taking these results as a confirmation of the straightforward theoretical deduction that, if a protein known by analysis not to contain an ideal proportion of amino-acids has those amino-acids in which it is lacking added to it, it will become as nutritious as casein, the President's Science Advisory Committee[34] made some calculations. For example, they calculated that at the 1967 prices of synthetic amino-acids, the cheapest way of bringing up the nutritional value of various cereal proteins to the value of milk casein would be to add these synthetic amino-acids to them in the proportions shown in Table 7.

Table 7

The cost of fortifying cereals to bring the nutritional value of their proteins up to that of casein with synthetic amino-acids, with fish-protein concentrate or with soya protein

Cereal	*Amino-acids required, %*	*Cost, cents/lb*	*Fish protein required, %*	*Cost, cents/lb*	*Soya protein required, %*	*Cost, cents/lb*
Millet/ sorghum	Lysine 0·45, Threonine 0·18	0·12	4·65	1·16	7·0	1·05
Wheat	Lysine 0·30 Threonine 0·15	0·53	3·10	0·77	4·7	0·71
Maize	Lysine 0·20 Tyrosine 0·35	0·27	2·05	0·51	3·1	0·47
Rice/ barley	Lysine 0·20 Threonine 0·10	0·35	2·05	0·51	3·1	0·47

Note: based on estimated costs of L-lysine monohydrochloride, L-threonine and L-tryptophan of $1, $1.50 and $2 per lb; and fish protein and soya protein of 25 and 15 cents per lb at 80% protein content.

These businesslike figures, and the tenor of the information presented in the whole of this chapter, show firstly, that it is not only possible to synthesise amino-acids, but also that amino-acids are being synthesised. Although they have not been employed on any large scale to the improvement of human diets, they have been used, on an entirely pragmatic and utilitarian basis, for animal feeding. But the fact that amino-acids can be synthesised, and at a comparatively low price that seems to possess an economic advantage over the cost of providing natural food, does not automatically imply that there is a nutritional need for synthetic amino-acids.

The diet of a large proportion of the world's population—that is to say, of the less fortunate proportion of mankind—is largely based on cereal grains and on cassava and other root crops.[38] It has been estimated that approximately 100 million people subsist mainly on cassava, while 1500 million people live on diets consisting predominantly of cereal grains.[39] It is among these people that protein deficiency is most frequently diagnosed, and among the children of the cassava eaters that the protein-deficiency disease, kwashiorkor, is most commonly found. Many of the mainly cereal diets eaten in developing countries, and even diets based on cassava as well, provide a reasonably adequate amount of total protein. It is generally accepted that a consumption of 1 g of protein per kg of body weight per day is amply sufficient. In predominantly wheat-eating communities, a wheat-based diet will supply 3·5 g of protein per kg of body weight. But the quality of the protein is poor, and whereas the average PER of wheat protein is 1·3, in some varieties of wheat it may only be 1·0. How low may it safely fall before trouble ensues? One answer is perhaps provided by Scrimshaw and his colleagues[40] who cured kwashiorkor by feeding a mixture of maize, sorghum and cottonseed meal with a PER of 1·8. Be that as it may, there is no doubt that the diet of a significant proportion of the less fortunate section of the world's population could be improved by enriching their natural diet with synthetic amino-acids which could at this moment be manufactured synthetically at a comparatively modest price. In Great Britain and the United States, to name but two advanced technological nations, the wealth of the State underwrites the supplementation of bread and of margarine with synthetic vitamins for the benefit of the less well-fed members of these lands. It is technically feasible to do likewise to the protein of the less well fed citizens of the world.

And if the researches in progress at the present time into the manufacture of "proteinoids" proceed in the way it is hoped, it may be feasible in due course to supplement the proteins which we have always so far obtained from biological sources by man-made 'flesh'.

[38] Brown, L. R. (1963), U.S. Dep. Agric. Foreign Ag. Econ. Rep., No. 11. Washington.

[39] U.S. Dep. Agric. (1962). *World Food Budget 1962 and 1966*, Foreign Ag. Econ. Rep., No. 4. Washington.

[40] Scrimshaw, N. S. (1961), Béhar, M., Wilson, D., Viteri, F., Arroyave, G., and Bressani, R., *Am. J. Clin. Nutr.*, 9, 196.

CHAPTER 3

SYNTHETIC FAT

Fat occupies a curious and somewhat ambiguous position in the dietary and nutritional scheme. As a dietary component it is prized. Indeed, the proportion of fat in the diet of a community can often be taken as an indication of the community's prosperity. Dry bread is a sign of poverty as well as being unattractive to eat. Bread and butter form an agreeable combination. The presence of fat is not the only important attribute of its dietary virtue; the nature of the fat is also of great social significance. In the industrialised parts of the world, store is placed on the firmness and solidity of fat, that is, upon its melting point. Butter and cooking fats are expected to be solid at normal climatic temperatures. One of the major applications of chemical science to food technology is the rationalisation of the ancient art of butter making to produce milk fat in an appropriate combination with water globules to constitute a firm solid. Another is the development of the process of hydrogenation. This allows any fat or oil to be 'hardened' to any appropriate degree, so that it may possess solidity and firmness under any climatic conditions.

Primarily, it could be assumed that fat is a luxury. While this is not entirely so, fat is, just the same, highly prized. Shortage of fat under wartime conditions is one of the most serious deprivations. Without fat, cooking becomes difficult and the variety of the diet is restricted. It was to meet a prognosticated fat shortage which, in the event, did not materialise, that the large-scale disastrous British project of the East African Groundnut Scheme was instituted.

Part of the dietary desire for fat is a reflection of nutritional need. Fat is the most compact source of nutritional energy. The calorific value of a gram of fat is 9, whereas that of a gram of either carbohydrate or protein is approximately 4. Further, since fat often occurs in food unmixed with water, its contribution of calories is, to that extent, increased. Heavy workers and other people who expend substantial amounts of energy in physical activity value fat, since it renders their food less bulky.

There are two other nutritional attributes of fat which give it value in the diet. First, it is a vehicle for the fat-soluble vitamins. For example, butter contains vitamin A and vitamin D. Fish-liver oils are particularly rich in these vitamins, although the oil is not very often used as a 'visible' fat in the diet. 'Visible' fats are such articles as butter, margarine, suet, cooking fats and cooking oils which are commonly obtained as ingredients in their own right. 'Invisible' fats are those which find entry into the diet as components of some other food. They contribute to the total fat content of the diet as it would be determined by chemical analysis, but can readily be overlooked by those who think about their food in a non-scientific manner. For example, the proportion of fat in spinach leaves or carrots is not very large, yet it may form a valuable ingredient, since it contains in it the quite considerable concentration of the vitamin A-active pigment, carotene. Fat also contains other valuable fat-soluble ingredients, including tocopherols (vitamin E) and vitamin K. But besides acting as a vehicle for these substances, fats may possess a positive nutritional attribute of their own. Although the evidence of their merits is not altogether clear cut, certain of the fatty acids that make up part of the molecular structure of some fats, the so-called 'essential fatty acids', may possess in themselves vitamin-like activity.

But the main attributes of fats as dietary ingredients are first, that they provide calories in conveniently compact form and, secondly, that people like to eat them. They like fried potatoes and fried fish, they prefer cake, and bread and butter, to bread alone, they enjoy cream in their coffee and oil on their salad. Fat is a prized article of diet. So much so, however, that affluent people may eat too much of it, and by doing so become obese, which tends to shorten their lives, or even contract coronary heart disease, and again die sooner than they otherwise would.

It is interesting to reflect that fat is, so far, the only food-stuff that has actually been manufactured by chemical synthesis on an industrial scale. It is true that, as is described in other chapters, certain vitamins and amino-acids are synthesised commercially and, as we shall discuss in Chapter 8, flavours, such as saccharine, and colours, are also manufactured synthetically, but these are all food ingredients. Fat is a food-stuff in its own right. And it was as long ago as the nineteenth century that the first experiments were carried out in which paraffins derived from petroleum were converted into long-chain fatty acids by air oxidation, and these subsequently esterified with glycerol to produce fat. Interest in this synthetic process was stimulated in Germany during the 1914–18 war by the food shortage brought about by the Allied blockade, but when the shortage of fat, which was the driving force behind such investigations, was relieved, interest quickly died down. Attention revived again, however, about 10 years later. This revival of scientific interest coincided with the development of a new starting material, the paraffin fraction called *Gatsch* of the Fischer–Tropsch process for making fuel oil and 'petroleum' out of coal. And so it happened, that before the Second World War began in 1939, German scientists were not only able to use coal, one of the main natural resources in which their country was richly endowed, as a source of strategic fuel for their military vehicles and aircraft, but they were also in a position to make edible fat out of it: a remarkable example of the use of scientific understanding to provide, as it were, both guns *and* butter.

Before the beginning of the Second World War, the Deutsche Fettsäurewerke at Witten had started producing synthetic fatty acids on an industrial scale. The fatty acids were used by the Chemische Fabrik Imhausen and Co. at Witten to develop production of edible synthetic fat. With commercial supplies of this synthetic fat at their disposal, physiologists and nutritionists were provided with a remarkable opportunity to study its wholesomeness and to investigate its metabolic behaviour in the body.[1]

Natural fats are mainly composed of triglycerides, that is, compounds of one glycerol molecule linked with three fatty-acid radicals. Usually at least two of the fatty acids are different. Besides triglycerides, fats may also contain a proportion of phospholipids, which are combinations of fats with phosphate. This phosphorylation, sometimes in combination with a nitrogenous base as well, is due to the biochemical elaboration in the body of substances concerned with special structures, such as cellular membranes. Sterols are another group of compounds commonly found in natural fats. In animal fats, the main sterol is cholesterol; in vegetable fats it is ergosterol. Sterols also serve particular biochemical functions. For example, as is described in Chapter 5, vitamin D is a member of the sterol group.

[1] Kraut, H. (1949), *Brit. J. Nutr.*, 3, 355.

When fat is considered as a food, attention must clearly be focused on the chemistry of its main triglyceride substance. The most abundant components of fats in nature are oleic and palmitic acids. The first of these comprises a chain of 18 carbon atoms and possesses one so-called 'double bond'; palmitic acid has a chain of 16 carbon atoms and, since it is lacking in double bonds, is described as 'saturated'. The formulae of these two fatty acids are:

$$\begin{array}{ccccccccccccccccccc}
 & \mathrm{H} & \mathrm{H} & \mathrm{H} & \mathrm{H} & \mathrm{H} & \mathrm{H} & \mathrm{H} & \mathrm{H} & \mathrm{H} & \mathrm{H} & \mathrm{H} & \mathrm{H} & \mathrm{H} & \mathrm{H} & \mathrm{H} & \mathrm{H} & \mathrm{H} & \\
 & | & | & | & | & | & | & | & | & | & | & | & | & | & | & | & | & | & \\
\mathrm{H}- & \mathrm{C}- & \mathrm{C}- & \mathrm{C}- & \mathrm{C}- & \mathrm{C}- & \mathrm{C}- & \mathrm{C}- & \mathrm{C}- & \mathrm{C}= & \mathrm{C}- & \mathrm{C}- & \mathrm{C}- & \mathrm{C}- & \mathrm{C}- & \mathrm{C}- & \mathrm{C}- & \mathrm{C}- & \mathrm{C}\begin{array}{l}\nearrow\mathrm{O}\\ \searrow\mathrm{OH}\end{array} \\
 & | & | & | & | & | & | & | & | & & & | & | & | & | & | & | & | & \\
 & \mathrm{H} & \mathrm{H} & \mathrm{H} & \mathrm{H} & \mathrm{H} & \mathrm{H} & \mathrm{H} & \mathrm{H} & & & \mathrm{H} & \mathrm{H} & \mathrm{H} & \mathrm{H} & \mathrm{H} & \mathrm{H} & \mathrm{H} &
\end{array}$$

Oleic acid

$$\begin{array}{ccccccccccccccccc}
 & \mathrm{H} & \mathrm{H} & \mathrm{H} & \mathrm{H} & \mathrm{H} & \mathrm{H} & \mathrm{H} & \mathrm{H} & \mathrm{H} & \mathrm{H} & \mathrm{H} & \mathrm{H} & \mathrm{H} & \mathrm{H} & \mathrm{H} & \\
 & | & | & | & | & | & | & | & | & | & | & | & | & | & | & | & \\
\mathrm{H}- & \mathrm{C}- & \mathrm{C}- & \mathrm{C}- & \mathrm{C}- & \mathrm{C}- & \mathrm{C}- & \mathrm{C}- & \mathrm{C}- & \mathrm{C}- & \mathrm{C}- & \mathrm{C}- & \mathrm{C}- & \mathrm{C}- & \mathrm{C}- & \mathrm{C}- & \mathrm{C}\begin{array}{l}\nearrow\mathrm{O}\\ \searrow\mathrm{OH}\end{array} \\
 & | & | & | & | & | & | & | & | & | & | & | & | & | & | & | & \\
 & \mathrm{H} & \mathrm{H} & \mathrm{H} & \mathrm{H} & \mathrm{H} & \mathrm{H} & \mathrm{H} & \mathrm{H} & \mathrm{H} & \mathrm{H} & \mathrm{H} & \mathrm{H} & \mathrm{H} & \mathrm{H} & \mathrm{H} &
\end{array}$$

Palmitic acid

There are several fatty acids that contain more than the one unsaturated double bond found in oleic acid. In arachidonic acid, for

Table 8

Proportion of the different fatty acids present in vegetable and animal fats, %

Fats	*Saturated fatty acids*				*Acids with one double bond*		*Acids with more than one double bond*		
Vegetable fats	C_{12} and less	C_{14}	C_{16}	C_{18}	C_{16}	C_{18}	C_{18} (2)	C_{18} (3)	C_{20} (4)
Coconut	60	18	10	2	0	8	1	0	
Palm kernel	61	18	7	2	0	11	1	0	
Olive	0	1	10	1	0	80	7	0	
Linseed	0	0	7	6	0	15	16	57	
Groundnut	0	0	8	4	1	53	26	0	
Maize	0	0	13	3	0	31	53	0	
Animal fats	C_{14} and less								
Beef fat	3		29	21	3	41	2	0	
Mutton fat	3		25	28	1	37	5	0	
Lard	1		30	16	3	41	7	0	
Whale oil	9		15	4	14	33	0	4	12
Butter fat	25		25	9	4	30	4	0	

example, there are four double bonds spaced along a chain of 20 carbon atoms. The proportion of unsaturated fatty acids in a particular fat is of importance, since the presence of unsaturation is associated with a lower melting point, fats richer in unsaturated acids will be liquid, that is, they will be *oils* at a temperature when more saturated fatty acids would lead to their being defined as fats. But besides unsaturation, the combination of the fatty acids in the triglyceride will also affect the character of the fat. Table 8 shows the proportion of fatty acids of different chain length present in different natural fats.

Two features of the chemical composition of natural triglyceride fats are, first, that in general the fatty acids possess an even number of carbon atoms in their chains and, secondly, the chains are straight not branched. The general chemical structure is of this form:

```
    H     O H H H
    |     ‖ | | |                                          Butyric acid radical
 H—C—O—C—C—C—C—H
    |       | | |
    |       H H H
    |
    |     O H H H H H H H H H H H H H H H
    |     ‖ | | | | | | | | | | | | | | |                  Palmitic acid radical
 H—C—O—C—C—C—C—C—C—C—C—C—C—C—C—C—C—C—C—H
    |       | | | | | | | | | | | | | | |
    |       H H H H H H H H H H H H H H H
    |
    |     O H H H H H H H H H H H H H H H H H
    |     ‖ | | | | | | | | | | | | | | | | |              Oleic acid radical
 H—C—O—C—C—C—C—C—C—C—C═C—C—C—C—C—C—C—C—C—H
    |       | | | | | | |     | | | | | | | |
    H       H H H H H H H     H H H H H H H H

Glycerol radical
```

The chemical distinction between edible oils—'oil' has no precise difference from 'fat'; 'oil' is merely a fat which happens to liquefy at the environmental temperature of the place where it is eaten—and mineral oils lie in the so-called carboxyl grouping, $-\mathrm{C}\begin{matrix}{}^{\nearrow}\mathrm{O}\\ {}_{\searrow}\mathrm{OH}\end{matrix}$, at the end of the carbon chain. Mineral oils are composed solely of carbon and hydrogen, and are, consequently, called *hydrocarbons*. Lacking in the terminal carboxyl group they are unavailable to the body and are, therefore, not food. In principle, however, the synthesis of fat must involve, first of all, the separation of hydrocarbon chains of the appropriate length from the mixture of which crude oil is composed before it has been refined. Refining mainly consists of separating by distillation the more volatile compounds used as petroleum in automobile and aircraft engines from the substances of larger molecular size constituting diesel fuel and heavier fuel oils, and freeing this fraction from the even less volatile fractions, which contain lubricating oil, petroleum jelly and, finally bitumen. As in the refining of edible fats, the process of oil refining also involves the separation of substances present in smaller concentration. These in crude oil will include sulphur-containing compounds and a variety of others, some of which give petroleum products their characteristic smell.

The large-scale manufacture of fatty acids, and hence of fats, achieved in Germany in the 1940s used coal as its basic raw material. The Fischer–Tropsch process that converted the coal into oil involved as a first step preparing a mixture of carbon monoxide and hydrogen by passing steam over white-hot coke. More hydrogen was added and the purified gas mixture then passed at a temperature of about 200°C over a catalyst prepared from cobalt and nickel. The effect of this was that carbon from the carbon monoxide (itself derived from the coal from which the coke was made) combined with the hydrogen to form hydrocarbon chains, some fully saturated and called *paraffins* and some partly unsaturated, chemically designated *olefins*. From the mixture, synthetic petrol called 'Kogasin' was separated, as well as diesel oil and other materials, including paraffin wax. The fraction boiling at between 320 and 450°C, called the Gatsch fraction, was used for the synthesis of fat.

This apparently simple process is a subtle and complex reaction in which the most delicate control of conditions is required. Its development was the result of a long series of trials and required highly sophisticated chemical understanding and considerable engineering ability. For example, it was only after 25 years research that in about 1923 the Badische Anilin und Sodafabrik first became able to operate a practical process. For this they used an iron catalyst over which they passed the gas mixture at high pressures of about 100 atmospheres. Although they obtained paraffin oils they also got a high proportion of oxidised products which they did not really want. Further work led to the development of improved iron catalysts which enabled them to produce a higher proportion of the kind of hydrocarbons they were aiming at and at a lower pressure of 1–15 atmospheres. It should be said that catalysts, which make these chemical reactions take place, are not even to this day fully understood, although numerous mathematical and physical theories have been elaborated to explain their action.

When the later cobalt and nickel and cobalt–thoria–magnesia catalysts were devised, control of the process was further improved and the proportion of paraffin and olefin hydrocarbons could be varied by appropriate adjustment of the gas mixture passed over them. Besides the Gatsch fraction used for subsequent fat synthesis, the Fischer–Tropsch process produced propane and butane, a gasoline fraction, a diesel oil fraction and solid paraffin wax. It is important to be aware of how complex is the mixture of materials, whether in the form of Gatsch or as a fraction separated at a petroleum refinery, from which the synthetic chemist sets out to manufacture artificial fat. It is not just fat that is required. The target is fat which will be acceptable alike for its chemical composition, its nutritional value and freedom from toxicity, and also for its taste and smell.

From 1884, when the first method for making fatty acids from hydrocarbons was proposed,[2] no serious attempt to manufacture them on an industrial scale was made until the end of the 1914–18 war when the price of crude whale oil—to cite it as an example—reached about £100 per ton. But the early processes, as I have already mentioned, gave mixtures of fatty acids together with various other oxidised products. An early solution to this problem was to oxidise only part of a batch of hydrocarbons, separate the fatty acids produced and then use the remainder over again as part of another batch. The fatty acids produced by this earlier process both in Germany and by Standard Oil at Baton Rouge in America possessed a very characteristic smell. A trace rubbed on to a man's hand left an odour which was difficult to remove.[3]

But the first real synthetic fat eaten was produced quite soon after the Markische Seifen Industrie in association with another soap manufacturer, Hankel et Cie, founded the Deutsche Fettsäurewerke at Witten in 1937. The process which these firms developed consisted of the following stages. First, batches of the selected Gatsch fraction were put into cylindrical aluminium vessels fitted with

[2] German Patent 32,705; British Patent 12,806 (1884).
[3] Williams, P. N. (1947), *Chem. & Ind.*, 251.

stainless-steel covers. Apparently the use of aluminium for the construction of the vessels facilitated the reaction. Another indication of the incomplete control of the process was the finding that a mixture of 8 tons of fresh Gatsch together with about 2 tons of Gatsch recovered from a previous run gave a better result than was obtained with only fresh Gatsch. To the starting material, about 0·5% of a 15% potassium permanganate catalyst mixture was added, the whole heated to 105°C, carefully mixed, and air blown into it through a system of very fine perforations.

The process of oxidation by air produces heat. Steps were therefore taken to hold the temperature at 105°C for 20–40 h, by which time about one-third of the original hydrocarbon was converted to fatty acid. This was washed out in hot water to remove the catalyst and unoxidised hydrocarbon and was then purified by being converted into soap with caustic soda to allow the removal of the last traces of unwanted material. The soap was then reconverted into the free fatty acids again by treatment with sulphuric acid, and the fatty acids were fractionally distilled at reduced pressure.

To convert fatty acids into fat, they must be combined with glycerol. This was carried out in 6-ton stainless-steel vessels fitted with water-cooled condensers. About 1% more fatty acid was used than was theoretically necessary and the fatty acid–glycerol mixture was heated to 120–180°C and kept at this temperature for 8 h in the presence of about 0·2% of tin or zinc which, this time, acted as a catalyst. At the end of the operation less than 1% of the fatty acids were left in a free form. The catalyst was washed out with a solution of dilute sulphuric acid, and the fat was then refined. Caustic soda solution was added to it and steam passed in. The caustic solution settled, carrying any free fatty acids with it. Fuller's earth was then added to the hot fat, as it has been for the same purpose since biblical times, and then charcoal. These absorb coloured materials and other impurities, which are removed when the fat is separated by filtration. The fat thus purified is then made into margarine.

Williams, in his account of the operation,[4] reported that most people commented unfavourably about the taste of the synthetic margarine, but accepted that it was not so bad that it would be refused by hungry men. On the favourable side of the balance sheet was its remarkably creamy consistency that made it easy to spread. It was

[4] Williams, P. N. (1947), *Chem. & Ind.*, 251.

also noticed that although the synthetic fat had a distinctly 'petroly' taste and smell, these were almost entirely absent from the margarine made from it. A second virtue of the synthetic margarine was that its keeping properties were exceptionally good. This was probably due to the absence of the traces of protein or other residues of milk which, while they may contribute a touch of flavour to butter, may also cause it to deteriorate if it is kept too long.

Table 9 shows one of the striking differences in composition between the synthetic fat and natural fats such as butter and coconut oil. It can be seen that whereas natural fats are made up solely of fatty acids containing even numbers of carbon atoms, the synthetic fat contained fatty acids with chains of both even and odd numbers of carbon atoms.

Table 9

The chemical composition of synthetic fat compared with that of butter fat and coconut oil,[5] %

Fatty acid	*Synthetic fat*	*Butter fat*	*Coconut oil*
C_8 and below	0	5·9	6·2
C_{10}	4·2	3·0	8·4
C_{11}	12·0	0	0
C_{12}	10·2	4·1	45·4
C_{13}	10·5	0	0
C_{14}	8·8	13·7	18·0
C_{15}	10·5	0	0
C_{16}	9·5	29·3	11·8
C_{17}	8·0	0	0
C_{18}	9·1	42·4	9·8
C_{19}, C_{20}	17·2	1·6	0·4

Although the flavour and consistency of synthetic fat and margarine made from it and their chemical composition as well are of importance, the nutritional value and freedom from toxicity are certainly of greater importance still. Studies on this point were reported by Flössner,[6] who was at one time head of the Physiological Department

[5] Williams, P. N. (1947), *Chem. & Ind.*, 251.

[6] Flössner, E. (1948), *Synthetische Fette, Beiträge zur Ernährugsphysiologie*, Ambrosius, Leipzig.

of the Reichs Gesundheitsamt. Feeding tests were carried out on thousands of mice, rats, guinea-pigs, rabbits and dogs. Some of these experiments were continued for five or six generations. The results of these trials showed that the synthetic fat was absorbed and was utilised by the body. At the end of the feeding period no pathological changes in any of the organs of the various types of animals were detected. These experiments, however, took no account of the observation made by several other experimenters[7] that the consumption of synthetic fat is followed by a very significant increase of the output of dicarboxylic acids in the urine. This aciduria shows that the metabolism of synthetic fat is undoubtedly different from that of natural fats.

More detailed consideration of some of Weitzel's results[8] led to less favourable conclusions. It appears that besides the normal straight-chained fatty acids, there were, in the synthetic fat, branched fatty acids as well as fatty acids containing hydroxyl groups, —OH, and keto-groups, $—\overset{\overset{\large O}{\|}}{C}—$, and in addition significant quantities of the dicarboxylic acid groups already mentioned containing two $—C\!\!\begin{smallmatrix}\nearrow\!\!O\\ \searrow\!OH\end{smallmatrix}$ groups instead of one as found in natural fatty acids. These variants, it appears are not readily assimilated or, if assimilated, are not metabolised efficiently and lead to the appearance of abnormal excretion products in the urine. In some of the experimental animals, growth was retarded. Goats fed on synthetic fat did not absorb it as readily as they absorbed butter. About one-third of the fat they laid down in their body tissues was found to be composed of unnatural fatty acids containing odd numbers of carbon atoms.[9] Although these fatty acids are abnormal, since they do not occur in nature, such evidence as there is[10] suggests that their nutritive value is similar to that of natural fatty acids containing even numbers of carbon atoms.

Although the excretion of increased quantities of dicarboxylic acid shows that the synthetic fat manufactured in Germany during the

[7] Thomas, K. (1946), and Weitzel, K., *Dtch. Med. Wschr.*, 71, 18; *Klin., Wochschr* (1949); Weitzel, K. *et al.* (1948), *Biochem. Z.*, 318, 472.

[8] F.I.A.T. Rep., No. 362, 3 (1945).

[9] Werner, K. (1939), Appel, H., and Boyer, G., *Z. Physiol. Chem.*, 257, 1; Emmerich, W., and Weber, M. (1941), *Z. Physiol. Chem.*, 266, 174.

[10] Appel, H. *et al.* (1943), *Z. Physiol. Chem.*, 274, 186.

Second World War was not metabolised in quite the same way as is natural fat, there is no evidence that it does not constitute a usable food. After having been tested on prisoners and on the workers at the Witten factory, the bulk of what was produced was supplied to the Wehrmacht and to the crews of submarines.[11] During the course of the war, four factories started producing synthetic fat in Germany, at Magdeburg, Heydebeck, Ludwigshafen-Oppau, and at Witten.

In his review of the wartime German achievement, Williams, writing in 1947, pointed out that when the technological feasibility of manufacturing fat from petroleum was considered in the 1920s, the price of natural fats ranged from £20 to £40 per ton, and it was concluded that artificial fat could not possibly be produced for such a price. During the war years of 1939–45, the estimated cost of producing synthetic fat was £177 per ton, so that only the extreme pressure of shortage made the operation worthwhile. After the stringency of wartime conditions passed, production was discontinued. It is clear, however, that synthetic fat can be manufactured and there is little doubt that should an economic demand again arise, the production of synthetic fat—of improved flavour and character now that more knowledge is available and the earlier experience can be used as a guide—could again be undertaken.

Although the synthetic fat manufactured from Gatsch possessed the nutritional disadvantage of containing a proportion of branched-chain fatty acids, it could, on the other hand, claim the advantage when compared with natural fat of not being a vehicle of sterols, and particularly of cholesterol. Cholesterol has been related to the incidence of coronary heart disease, particularly among middle-aged men who eat too much and too rich a diet and take too little exercise. Apart from providing a concentrated source of calories, making the diet more palatable and contributing to its culinary quality, fat delays the passage of food through the digestive tract, and consequently prolongs the sensation of satiety after a meal. In addition, however, the chemically 'saturated' long-chain fatty acids have been found to exert a tendency to raise the concentration of cholesterol in the blood-stream. In contrast, 'unsaturated' fatty acids, of which linoleic, linolenic and arachidonic are the most important, exhibit a tendency to reduce the concentration of cholesterol. Such unsaturated compounds are sometimes referred to as 'essential fatty

[11] F.I.A.T. Rep., No. 364, 16 (1945).

acids', and claims have been made for their inclusion in the list of vitamins on account of this and some other activity.

Although the wartime process for synthesising fat in Germany was in the main designed to produce saturated fatty acids from paraffins —that is, from chemically saturated hydrocarbons—fatty acids may also be produced from olefins as well—that is, from hydrocarbons containing one or more unsaturated double bonds. A procedure known as the Oxo process was developed in Germany for producing fatty acids from olefins, and had been developed up to a scale of 10 000 tons per year by 1939. By using such a procedure, synthetic fat containing any desired proportion of unsaturated fatty acids could be produced. At the time the method was elaborated, the physiological significance of the unsaturated, so-called 'essential', fatty acids had not been recognised.

In the Oxo process, a mixture of unsaturated fatty acids is obtained by 'cracking' a hydrocarbon fraction containing paraffins with chain lengths varying from 20 to 40 carbon atoms. Alternatively, a more precisely selected group of olefins can be produced directly by a careful modification of the Fischer–Tropsch process. Of course, just as specific vitamins can be synthesised as pure chemicals free from other contaminating substances, so also can oleic acid, $CH_3(CH_2)_7CH{=}CH(CH_2)_7COOH$, be synthesised.[12] Methods for synthesising linoleic acid, $CH_3(CH_2)_4CH{=}CH \cdot CH_2 \cdot CH{=}CH(CH_2)_7COOH$, have also been published.[13] These methods have not been used on a large scale, but they show that a tailor-made fat could be produced if the demand for its manufacture was sufficiently pressing.

What I have described so far shows that the manufacture of synthetic fat, as a purely artificial food prepared from inedible raw material, is entirely within the bounds of scientific feasibility. This synthesis was, in fact, carried out under the exceptional economic stresses to which Germany, a country of great technological sophistication, was exposed in wartime. But chemical synthesis making available substantial tonnages of fat for human food is in operation at the present time. I refer to the phenomenal upsurge in the production of synthetic detergents which releases for eating the great quantities of fats which, since times of antiquity, had otherwise been used to make soap. In 1945, approximately 140 million lb of synthetic

[12] Robinson, G. M. (1925), and Robinson, R., *J. Chem. Soc.*, 127, 175.
[13] Noller, C. R. (1937), and Girvin, M. D., *J. Am. Chem. Soc.*, 59, 606.

detergents were used in the United States. Five years later, the production of synthetic detergents was 1100 million lb per year. By 1955, the figure had reached 2320 million lb, at which point it began to level out so that by 1960 it had become about 3000 million lb. While this extraordinary growth of chemical manufacture was taking place—twenty-fold in 15 years, an increase paralleled in other industrialised countries—the sale of soap in the United States fell from about 3500 million lb in 1947 to 2000 million lb in 1953.[14,15]

To make soap, the combination of fatty acids and glycerol of which a molecule of fat is composed must be broken. The glycerol is separated and the fatty acids combined with a sodium or potassium ion derived from the caustic alkali which is used in the manufacture. To make a ton of soap, something of the order of a ton of fat is needed. The release of fat due to the sudden increase in the use of synthetic detergents was a remarkable historical event, exerting an effect on the world's available food supplies almost equivalent to the discovery of a new continent. If, in 1944, anyone had predicted that the quantity of synthetic detergents used in the United States would become greater than that of soap, the proposition would have been dismissed out of hand. Yet within 10 years this had occurred. In the main, this was due to the development of the so-called alkylarylsulphonates. Part of the dominant position of these compounds, which are made from petroleum derivatives, is due to their general effectiveness as washing agents. But a very important part of their commerical success is also due to two economic factors: first, the virtually unlimited supplies of uniform raw material, mainly polypropylene-benzene, and, secondly to the fact that the price of this raw material has been kept substantially stable. Attempts to manufacture food from unusual sources, for example protein from leaves and fish flour, or yeast from surplus molasses, have been bedevilled by commercial failure, even after technical manufacturing processes of great skill and insight have been developed. These failures have often been due to fluctuation in the supply and in the price of the raw materials needed.

A typical structure of the raw material from which synthetic detergents are made, tetrapropylenebenzene, is shown opposite:

[14] Fechor, W. S. (1959), Strain, B., Theoharus, L., and Whyte, D. D., *Ind. Eng. Chem.*, 51, 13.

[15] Bramston-Cook, H. E. (1954), and Elwell, W. E., *Ind. Eng. Chem.*, 46, 1922.

```
        H                    H
        |                    |
     H—C—H                H—C—H
   H    |       H  H     H   |   H
   |    |       |  |     |   |   |                     H
H—C—C———C—C———C—C—C————C=C<
   |    |       |  |     |   |   |        |            H
   H  H       H  |     H  H  H        |
                H—C—H                      |
                   |                 H—C⸗C⸌C—H
                   H                     |        ||
                                     H—C⸌C⸗C—H
                                            |
                                            H
```

This basic structure (which is not a rigidly defined chemical substance and of which there are several known variants), forms the basis, when sulphonated, of an excellent detergent. It possesses, however, one drawback, whose solution had led to a rather peculiar scientific paradox. Detergents made from chemical intermediaries of this sort cannot readily be broken down biologically; consequently they are a nuisance to the people who operate sewage plants, as well as to the general public, who do not welcome their lakes, ponds and streams being blanketed by permanent and almost unbreakable billows of indestructible foam. It is, therefore, interesting to note that in devising a way to produce biologically 'soft' detergents capable of being degraded by the micro-organisms used in sewage works, the chemists have adopted the principle of producing as basic raw material straight-chain rather than the branched-chain intermediaries, whose configuration is set out above.[16] Whereas propylene polymers are readily and conveniently produced during the refining of high-octane fuel, straight-chain olefins could at first only be produced on a commercial scale from the 'cracking' of petroleum waxes. 'Cracking' is the process whereby the chain length of the compound is shortened. Even so, a proportion of branched-chain material remained in the product, so that the resultant detergent was only partly degradable. A new and more completely degradable detergent has since been developed, in which alcohols rather than olefins are used as a basis. These are produced by the petrochemical process of cracking using a special Ziegler catalyst.

[16] Stupel, H. (1964), *Chem. & Ind.*, 470.

These sophisticated catalysts, for whose development Professor Ziegler was awarded a Nobel Prize, induce what is called an 'Aufbau' synthesis. This starts with the production of triethylaluminium, which forms the basis upon which more and more 2-carbon-atom ethylene molecules are built up to form long carbon chains. Next, these are oxidised in air under pressure; water molecules are then introduced to produce alcohols.

The curious and paradoxical feature of this development aimed at producing 'soft' synthetic detergents that can be degraded by the biological action occurring in a sewage works is that, to achieve this result, the non-edible polyalkyls from which synthetic detergents were first made are gradually being modified—by purely chemical means—to produce alcohols which are, to a greater or lesser degree, edible by people as well as by sewage micro-organisms. It would be a curious, but, in its way, satisfactory, outcome if, in pursuit of a synthetic detergent that would be effective as a washing agent and at the same time readily degradable after it had been used, the synthetic chemists eventually modified their product and then modified it again until in the end they had invented—soap. But this time it would be soap, not made from fat which might otherwise have been eaten, but from synthetic fat. Problems of flavour, metabolisability and the possibility of toxicity would then, in one stroke, have become irrelevant.

But the discovery of organo-metallic compounds as catalysts, of which those associated with the name of Ziegler were the first and the most dramatic,[17] opened up possibilities for the synthesis of fat which have not so far been exploited. The discovery by Ziegler that the 2-carbon-atom molecule of the olefin, ethylene, $H{-}\overset{H}{\overset{|}{C}}{=}\overset{H}{\overset{|}{C}}{-}H$, could be polymerised under less extreme conditions than was otherwise possible provided triethylaluminium or diethylberyllium were present as catalysts, was first used in the preparation of polyethylene. In due course it emerged that the catalyst operated on a molecular scale much as if it were the slide of a zip-fastener, by means of which the two parts which enter separately emerge linked together. An even more refined organo-metallic catalyst was subsequently developed

[17] Ziegler, K., and Gellert, H. G., German Patent 883,067; Ziegler, K. (1955), Holzkamp, E., Breil, H., and Martin, H., *Angew. Chem.* 67, 541.

by Natta in Italy[18] by which the orientation in relation to each other of the molecules which were joined could be controlled. These spectacular advances in applied organic chemistry open up the possibility of linking up simple 2- and 3-carbon-atom compounds, such as ethylene and propylene, into chains of controlled length and conformation. Consequently, just as fatty acids, and hence fats, can be made by suitably oxidising existing long-chain hydrocarbons, so does this new knowledge open the possibility of linking together the kinds of simple intermediaries which are commonly produced in the petrochemical industry and which also occur as major components of natural gas.

As with the other foods and nutrients that can also be synthesised in chemical factories from non-edible raw materials, fat can quite readily be made by synthetic means. Whether or not this is done depends, as it does for other foods, on the synthesis being economic. The assessment of whether or not this is so is in some ways particularly difficult for fat. For example, although certain fats—for example, high-quality olive oil—possess special properties upon which their individual price depends, for the most part, one fat can be turned into another almost at will. Solid fats can be turned into oil, that is to say, their melting point can be lowered by 'cracking'. On the other hand, liquid fats—that is, oils—can be hardened by hydrogenation. Between 1900 and 1920, the consumption of margarine in the United States increased from 0·5 to 3·4 lb per head of the population, and the consumption of compound cooking fat rose from 4·5 to 11·7 lb per head.[19] Both of these commodities are made indifferently from those fats most readily available on the market. Depending on supply—which affects availability and price—whale oil, other animal fats, vegetable oils, such as peanut, soya bean, coconut or palm kernel oils, or fish oils are put through the highly efficient industrialised processes of deodorisation and refining and are then to a large extent brought to any chemical composition required. That is to say, the lengths of the respective fatty-acid chains can effectually be adjusted and the degree of unsaturation brought to any desired value. It follows, therefore, that if the cost of, say, soya bean oil rises, it can be replaced by fish oil or cottonseed oil.

[18] Natta, G. (1955), *Macromol. Chem.*, 16, 213; — (1955a), *J. Polym. Sci.*, 16, 143.

[19] Bailey, A. E. (1945), *Industrial Oil and Fat Products*. Interscience, N.Y.

The cost and availability of fats for food may not only be influenced by factors affecting the food industry alone or the nutritional requirements of a country. I have already referred to the competing demand for soap. The usage of soap has in the past often served as an indicator of the standard of living of a developing nation. This index still applies, even if to a more limited degree on account of the rapid and dramatic increase in the manufacture of detergents mainly synthesised from petroleum chemicals. It is interesting to reflect on the parallel revolution that occurred 50 years earlier, when the substantial diversion of potentially edible fat for use as an illuminant in lamps ceased, first, due to the use of kerosene (so-called 'paraffin') derived from petroleum, later, by the introduction of gas light and, subsequently, of electric lighting. The diversion of significant quantities of oils and fats as an ingredient of paint has also diminished with the advances in paint chemistry.

Finally, in reflecting on the economic feasibility of the large-scale chemical synthesis of fat, we must take into account the very extensive research which has been carried out in an effort to produce fat by propagating micro-organisms. Numerous such biosynthetic methods have been reported,[20] and among the micro-organisms employed have been *Endomycopsis vernalis*, *Oidium lactis*, *Penicillium*, *Aspergillus*, *Mucor*, *Fusarium* and both *Torula* and *Saccharomyces* yeasts. Some of the processes developed have been operated, if not on a full scale, at least on a large pilot-plant scale. In spite of all the intensive research that has been carried out, however, the cost of the fat so produced has been very much higher than that of natural fat.

Fat is prized as a component of diet. Although there are marked differences in the amount of fat eaten by different communities, it is a valued commodity, of which at least a small proportion is considered necessary for proper nutrition, quite apart from its aesthetic and culinary contribution. There is little doubt that the scientific and technical knowledge is available to synthesise fat from hydrocarbons derived either from coal, petroleum or natural gas. Whether this can profitably be done must depend, in large measure, on economic circumstances.

[20] Hesse, A. (1949), *Adv. Enzymol.*, 9, 653.

CHAPTER 4

SYNTHETIC CARBOHYDRATE

Few chemists, and even fewer nutritionists, are aware that the synthesis of carbohydrate was first reported almost 90 years ago. In 1861, Butlerow[1] observed the formation of sugar-like substances when formaldehyde was treated with mild alkali. Since that time, the mechanism of the reactions involved has been studied by several workers.[2,3] Mayer and Jäschke, working at the Technische Hochschule in Dresden, have shown[4] that, under appropriate conditions, the "formose reaction" can be made to yield the mixed sugars, glucose, galactose, arabinose and xylose. Fructose, the most prominent component of honey, was first identified by E. Fischer[5] as a component of "formose" as long ago as 1888.

The synthesis of hexose sugars, which are the basic components of food carbohydrates, from so simple a starting material as formaldehyde is of the highest importance, and it is surprising that almost no work was done on the reaction between the mid-nineteenth century, when the formose reaction was discovered, and Mayer and Jäschke's studies of 1960. If serious development work should show—as has happened with so many other seemingly impractical reactions before—that efficient yields could be obtained, then indeed would the synthesis of human food have become a practical proposition.

The main food carbohydrates are made up of units of 6-carbon-atom sugars, of which the commonest are glucose and fructose. The chemical configuration of these is shown below:

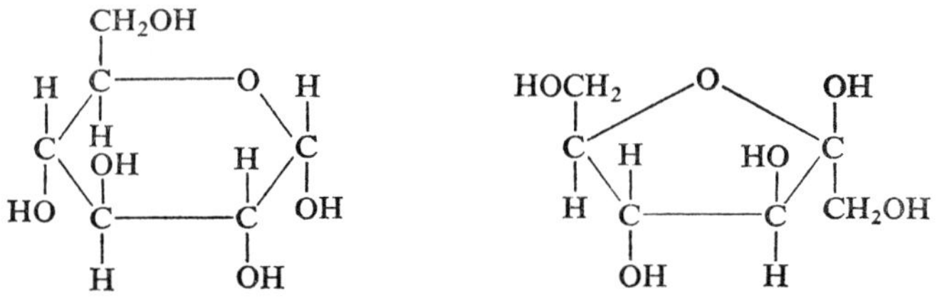

CH_2OH | H C—O H | H OH H C | HO C—C OH | H OH

Glucose

$HOCH_2$ O OH | C H HO C | H C—C CH_2OH | OH H

Fructose

[1] Butlerow, A. (1861), *C.R. Acad. Sci., Paris*, 53, 145.
[2] Langenbeck, W. (1958), *Tetrahedron*, 3, 185; — (1954), *Angew. Chem.*, 66, 151.
[3] Breslow, R. (1959), *Tetrahedron Letters*, No. 21.
[4] Mayer, R. (1960), and Jäschke, L., *Ann.*, 635, 145.
[5] Fischer, E., *Ber.*, 21, 988 (1888).

Sucrose, known in common speech as 'sugar', is a disaccharide comprising two linked hexose units, one glucose and one fructose. Its configuration is shown below:

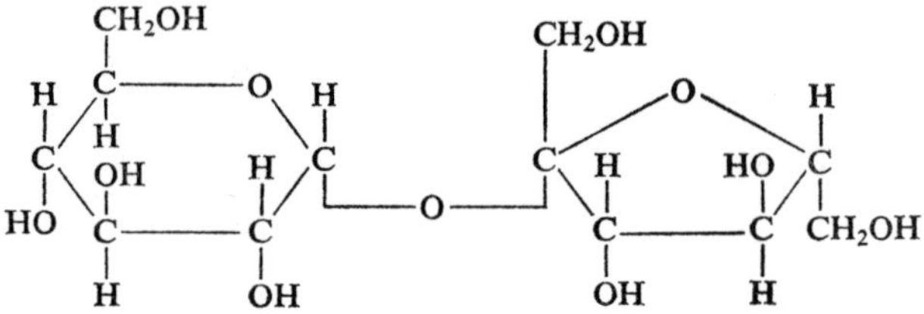

```
          CH2OH
           |                              CH2OH
    H      C----------O   H                 |
    |     /|           \  |                 |     .O.            H
    |   /  H            \ |                 |   /     \          |
    C       OH       H   C                  C    H      HO  C
    | \     |        |  / |                 |\   |       | / |
   HO   \   C--------C    L------O----------J \  |       |/  |
            |        |                          C--------C  CH2OH
            |        |                          |        |
            H        OH                         OH       H
```

Sucrose

The separate chemical synthesis of glucose and of fructose was first carried out in 1887 by Emil Fischer and his colleagues.[6] The starting material was again formaldehyde treated with alkali which, as I have already stated, leads to the production of a mixture of sugar-like compounds. Fischer, however, carefully separated and purified one of these, which he identified as a compound of the 5-carbon-atom sugar, arabinose. With this as a basis, he and his co-workers developed a process requiring four separate stages for the production of fructose and a further four to obtain glucose.

In view of these early successes, and of the fact that the raw material, formaldehyde, is a common chemical derived comparatively cheaply from petroleum or coal, it seems surprising that little practical work has been done to make use of these discoveries. Further investigations have, of course, been carried out. For instance, in 1953, two Italian chemists,[7] by making use of modern analytical methods unavailable in Emil Fischer's time, identified glucose and fructose, among other sugars, including galactose, sorbose, mannose, arabinose, xylose, lyxose and ribose, from a formose preparation made by heating formaldehyde under certain specified conditions.

Formaldehyde, $H{-}\overset{\overset{O}{\|}}{C}{-}H$, is not the only substance which, when heated with alkali, yields sugars. Emil Fischer himself showed that similar reactions occur when glycerol which, as was described in the

[6] Fischer, E. (1887), and Tafel, J., *Ber.*, 20, 2566; Fischer, E. (1890), *Ber.*, 23, 370, 799.

[7] Mariani, E. (1953), and Torraca, G., *Intern. Sugar J.*, 55, 309.

last chapter, can itself be obtained from hydrocarbons, is oxidised to a mixture of glyceraldehyde,

```
     OH   OH   O
     |    |    ||
  H—C————C————C
     |    |    |
     H    H    H
```

and dihydroxyacetone

```
    OH    O  OH
    |     ||  |
 H—C————C—C—H
    |        |
    H        H
```

and these are then heated under alkaline conditions. There are, indeed, several compounds that react, although in rather an uncontrolled manner, to yield a mixture of further compounds, among which are sugars of various kinds. Actually, the substance used originally by Butlerow was not pure formaldehyde, which was not available in his day. Instead, it was a mixture of polymerised formaldehyde compounds which occur when formaldehyde is allowed to stand exposed to air. This so-called 'paraformaldehyde' comprises substances with molecular structures in the form of chains of formaldehyde units of varying length with the general configuration:

```
    OH     H     H     H          OH
    |      |     |     |          |
 H—C—O—C—O—C—O—C—O· · ·—C—H
    |      |     |     |          |
    H      H     H     H          H
```

The exact way in which the formose reaction is carried out, that is to say, the alkali used, the length of time the reaction is continued and the precise starting material used, has been shown to affect the mixture of compounds that is produced. Schmidt,[8] for example, succeeded in obtaining fructose and sorbose. This same achievement was repeated, but in a more precise way, 20 years later.[9] It seems clear that the variety of sugars which may be produced is extensive. A

[8] Schmidt, E. (1913), *Ber.*, 46, 2327.
[9] Fischer, H. O. L. (1936), and Baer, E., *Helv. chim. Acta.*, 19, 519.

curious sugar called 'dendro-ketose' with a branched molecular chain structure has even been identified.[10] Its configuration is:

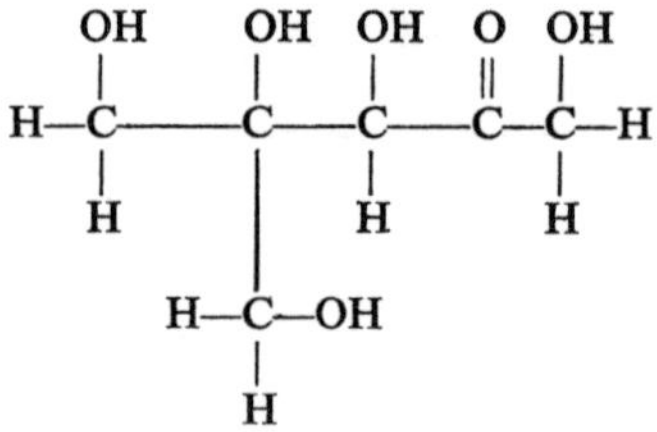

```
   OH       OH   OH   O   OH
   |        |    |    ||  |
H—C—————C————C————C—C—H
   |        |    |        |
   H        |    H        H
            |
         H—C—OH
            |
            H
```

Dendro-ketose

Glucose and fructose are merely two of the commonest sugars which occur in nature. In addition to these there are 5-carbon-atom pentoses. Artichokes are examples of plants producing pentoses. Clearly, small differences in conditions influence the exact structure of the sugar produced. It follows, therefore, that in the artificial circumstances of the formose reaction, which are inevitably less subtle than those of biosynthesis in plants, a variable mixture of compounds is produced.

Carbohydrates, the main energy-giving components of food, are all based on sugar units. But although it is a remarkable achievement to have succeeded in synthesising simple sugars in the laboratory, either singly, as was done by Fischer, or as occurs in the formose reaction, as a mixture of diverse sugars, this does not solve the complete problem of making 'bread' artificially. Even the detailed conditions under which an appropriate variant of the formose reaction gives an optimal yield has not yet been worked out. Carbohydrates, in the form in which they are mainly eaten, are not single sugars. Glucose, for example, although the component into which starch is broken down in the body, is only rarely eaten as such. It is, therefore, necessary to consider the synthesis of compound sugars.

Sucrose, a combination of glucose and fructose combined by the so-called 'α-*d*-glucosyl' linkage, is a major article of trade. It is, in fact, almost the only significant food-stuff that is a single pure chemical compound. While comparatively common in nature, its chemical synthesis, which has been successfully achieved, has proved quite difficult.

[10] Utkin, L. M. (1949), *Doklady Akad. Nauk. S.S.S.R.*, 67, 301.

The method which was at last successfully worked out by two Canadian scientists,[11] was exceedingly complex. First of all a carefully prepared preparation of glucose had to be made. The glucose molecule was combined in an exactly appropriate way with acetyl radicals in order to ensure that when the linkage with fructose took place it would occur at the proper position in the molecule. The same careful combination of fructose with four acetyl radicals also appropriately positioned had also to be prepared. These were then mixed together and heated for 104 h in a sealed tube at 100°C. After this, the product had to be de-acetylated and the separate members of the many compounds formed during the course of the heating separated one from another. This was done by passing the solution in which they were dissolved through a chromatographic column. Eventually, 5·5% of the original glucose and fructose was recovered in the form of sucrose. Although this represented a scientific *tour de force* and a milestone in the progress of organic chemistry, it was in no sense a practical process for making sugar. Nevertheless it is from such beginnings that great developments have emerged. The separate sugar units, glucose and fructose, can, it appears, be synthesised by some appropriate variant of the simple formose reaction. The combination of fructose and glucose is not particularly difficult, but the problem is to cause them to combine by the appropriate linkage. Lemieux and Huber succeeded in doing this by blocking off all the alternative ways by which the molecules might otherwise have joined. There is, however, another possibility. It is, as I shall describe in a moment, to use an appropriate catalyst.

Sucrose, although it is a useful and agreeable food-stuff, is not of paramount importance. Starch is a far more important compound. It is composed, not as sucrose is, merely of two hexose units linked together, but of long chains of glucose units. When the chains are unbranched, the starch is called 'amylose'; when they are branched, it is called 'amylopectin'. As a general rule, the starches with which we are most familiar in such foods as bread, rice or 'custard powder' (maize starch) are about three-quarters amylopectin and one-quarter amylose. The first laboratory synthesis of starch was carried out in 1940.[12]

[11] Lemieux, R. U. (1953), and Huber, G., *J. Am. Chem. Soc.*, 75, 4118.
[12] Hanes, C. S. (1940), *Proc. Roy. Soc. Lond., B.*, 128, 421.

Hanes, who achieved the synthesis of starch in the laboratory, had great difficulty in doing so. Indeed, he only succeeded by using purified enzymes which can be isolated from plants or from micro-organisms. These enzymes are themselves complex chemical compounds which serve as catalysts to bring about only the single chemical reaction for which each one is specific. Hence, although Hanes's synthesis of starch was a non-biological operation, he did use reagents, namely enzymes, which were themselves of biological origin. In order to get the individual glucose molecules to link together, he found it necessary first to convert them into an 'activated' form by combining them with phosphate. Only then could they be linked into the appropriate chain formation. The sequence of events is as follows.

The simplest form of activated glucose is the phosphate (*α-d*-glucose-1-phosphate):

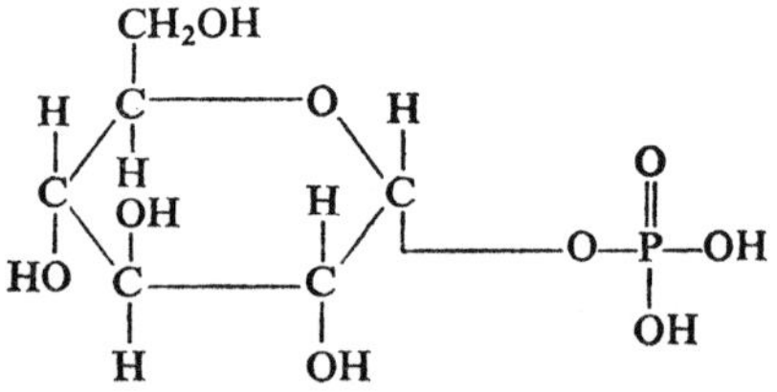

```
           CH2OH
            |
      H     C---------O   H
      |   / |          \  |                 O
       C    H            C                  ||
      |  \  OH        H /  \_________O--P--OH
     HO   C-------------C               |
           |            |               OH
           H            OH
```

This was formed with the aid of the appropriate enzyme isolated from plant tissue. This compound can itself take part in polymerisation or it may be converted into two more complex compounds that also operate as donors of the glucose units to be 'knitted' together one by one to produce the chains constituting the starch. One of these more complex compounds is uridine diphosphoglucose:

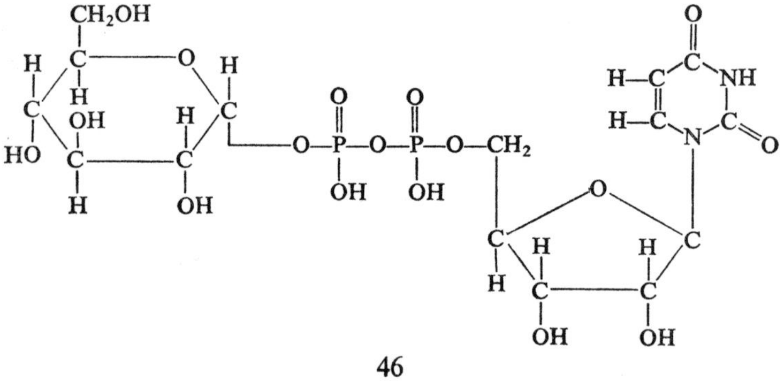

```
       CH2OH                                                   O
        |                                                      ||
   H    C---------O   H                                   H--C--C--NH
   |  / |          \  |       O     O                        ||    |
    C   H            C        ||    ||                    H--C--N--C=O
   | \  OH        H /  \___O--P--O--P--O--CH2                   |
  HO   C-------------C        |     |      |                    |
       |             |        OH    OH     |       O            |
       H             OH                    |   /       \        |
                                           C   H      H   C
                                           |   |      |  /
                                           H   C-------C
                                               |       |
                                               OH      OH
```

and the other is adenosine diphosphoglucose:

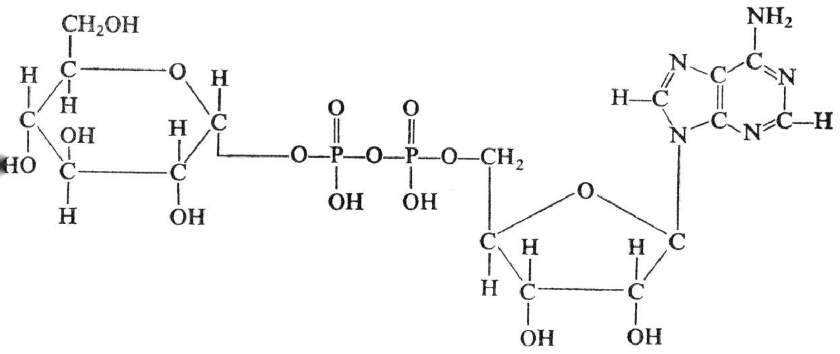

```
                                                                              NH2
      CH2OH                                                                    |
       |                                                                  N—C—C=N
 H     C———————O   H                                                   H—C   ||  |
 |   / |        \  |         O      O                                    N—C—N=C—H
 C     H         \ C         ||     ||                                   |
 | \  OH   H    /  |_____O—P—O—P—O—CH2                                   |
HO   C————————C              |      |      |           O                 |
     |        |              OH     OH     |      /         \            |
     H        OH                           C   H           H   C
                                           | \ |           |  /
                                           H   C———————————C
                                               |           |
                                               OH          OH
```

Again, before Hanes was able to make either one of these, he first had to purify the appropriate enzyme.

The fourth form of activated glucose from which the finished starch can be prepared is a length of polymer, made up from as few as two glucose units or as many as several hundreds.

Obviously, the delicate and complex procedures which Hanes had to elaborate in order to achieve starch synthesis in the laboratory are highly unlikely to form a practical manoeuvre for synthesising starch. The reason why starch synthesis, piece by piece like this, is complicated is explicable when one considers the diagram below. Starch, the main component of bread and the main source of the biological energy of higher animals, possesses the same empirical formula as cellulose, one of the structural components of wood from which paper and cotton—both largely inedible by man and other non-ruminants—is made up. Both are composed of chains of glucose units. But because the units in starch are all spatially the same way up it is edible, and because the units in cellulose are linked one up, the next down, and so on, it is inedible. The problem of synthesising starch is, therefore, not merely to link the glucose units together, but to link them together all the right way up in space.

The complex nature of the compounds constituting the activated form of glucose is, therefore, explicable when one considers the biological importance of the exact molecular configuration of the glucose polymer that finally emerges. When we appreciate that the exact 'twist' of the molecule decides whether the polymer is to be starch—digestible and metabolisable—or whether it is to be cellulose

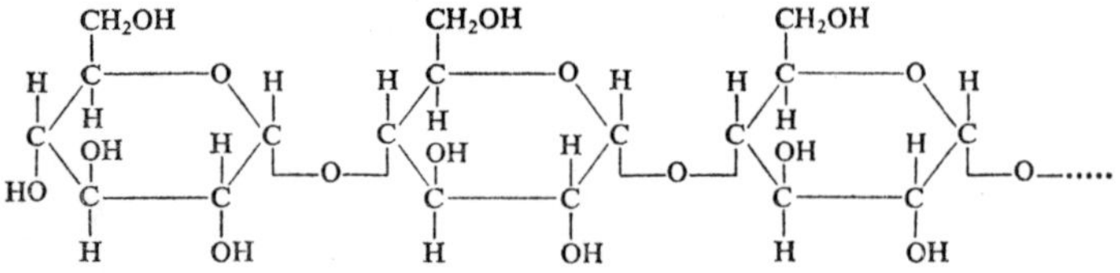

Starch fragment

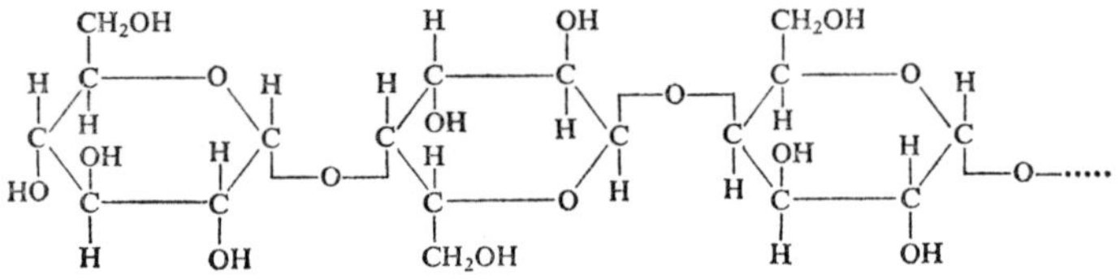

Cellulose fragment

—tough, rigid and resistant to biological breakdown—the precise functioning of the glucose donors (i.e. how they add each glucose unit to the chain which is eventually to become the fully synthesised polymer) is readily apparent. In the biological processes of life, the starch molecule is broken down to render up its content of energy through the intermediacy of the series of enzymes we human animals possess. Each of these is quite specific for its purpose. Such an enzyme can be compared with the slide of a zip-fastener. This slide fits the starch polymer and releases glucose units from the chain so that we can utilise them for energy. It does not fit the cellulose polymer.

The relevance of the exact molecular arrangement of a starch molecule is seen in the visible structure of starch under the microscope. If one makes a molecular model out of wood and wire, based on known molecular dimensions, it will be found that as the chain of glucose units lengthens it takes the form of a spiral or helix when the known configuration of starch is followed. On the other hand, if the glucose units are put together in the way that they are known to be joined in cellulose, that is, alternately facing 'up' and 'down', the chain as it grows remains straight. These molecular conformations are reflected in what can be observed of the material structure of starch and cellulose. Starch occurs in the form of granules. The twisted molecular chain winds round itself like a ball of wool, and it is from such molecular balls that the grains develop. In the plant where the biosynthesis occurs, starch constitutes stored fuel to maintain life processes until newly grown leaves, spreading their green chlorophyll to catch the energy of light, produce a fresh supply of sugars. Used for human food, the configuration of the starch can readily be broken down to release its energy in the human body. The comparatively small change in molecular configuration represented by the linkage in the cellulose chain gives rise to a straight molecule from which strong fibres of structural substance are built up. These fibres are unavailable as a source of biological energy to higher animals. Only cows and other ruminants can use the chemical molecule of cellulose, and they do this only through the intermediacy of lower organisms, namely the micro-flora of their rumen.

Threading glucose units together in the orthodox way in which Hanes succeeded in doing it—by using complicated 'activated' forms of glucose compounds, each of which can only be produced by its appropriate enzyme—seems to be a hopeless method of manufacture. It may, indeed, become a practical proposition if the hopes of those people who believe that enzyme technology is likely to become a matter of practical economic importance in the future are realised. There is, however, an alternative possibility that is at least worthy of mention. During recent years, there have been major developments in plastics technology due to the development of so-called 'organo-metallic' catalysts. These remarkable substances, for which Ziegler in Germany and Natta in Italy were awarded Nobel Prizes, were responsible for the radical simplification in the manufacture of polyethylene and polypropylene, both of which, though simpler than starch, are nevertheless also long-chain structures. It is, therefore, worth speculating as to whether at some time in the future a more advanced type of purely artificial organo-metallic catalyst may be developed to take the place of Hanes's enzymes and activated forms of glucose and allow starch—or for that matter cellulose—to be synthesised in bulk. This would enable a proportion of our food starch—and cotton as well, perhaps—to be 'man made'.

The controlled synthesis of starch, though possible, is obviously very difficult. But it may not be necessary. Just as the still more difficult problem of synthesising protein may well have been circumvented by the discovery of 'pansynthesis', so also may a number of reports in the scientific literature indicate a short cut to the synthesis of starch; or if not starch, at least something like it.

When bread is toasted, the molecules of starch are degraded and a mixture of substances of varying molecular size is produced. These substances are known collectively as *dextrins*. Although they do not possess the fine molecular structure of starch they constitute nevertheless a nutritious and agreeable component of the diet. Unless, that is, the treatment by heat has been carried so far that the molecular integrity of the glucose polymer has been completely disrupted and the toast has been burnt. This being so, some interesting synthetic possibilities arise, providing we are satisfied to make something less than starch.

By taking formaldehyde from a gas works or a petrochemical plant and heating it with alkali we can, according to Butlerow and the other authorities to whom I have already referred (see footnotes 1–5, p. 41) produce a mixture of sugars. If we now purify the mixture, separate the *d*-glucose from the other sugars and heat it with phosphoric acid, a polymer which is a kind of dextrin will be produced. This reaction was discovered as long ago as 1858 by Berthelot.[13] Another early discovery which, like the first, also clearly warrants further detailed investigation, was that of Musculus,[14] who found that if glucose is treated with concentrated sulphuric acid a white, sticky compound is obtained chemically similar to dextrin made from starch. Many other chemists have made similar observations and the process of 'reversion', as it is called (i.e., the tendency of individual sugars to link together in the presence of acid), is recognised as one way in which condensation polymerisation takes place and polysaccharides are produced. But not only can separate sugars be made to combine together in this way; groups of already-linked sugar molecules (sometimes called oligosaccharides) can themselves be caused to link together to form even more closely starch-like polymers by this process.[15]

A very much higher degree of control over the kind of linkage obtained in the production of these artificial dextrins has been achieved by G. Schramm and his colleagues.[16] They did this by using a special catalyst, namely polyphosphoric acid ester, to activate the linkage of glucose and other sugars. Good yields of *d*-glucose polymers with a molecular weight of about 50 000 and of polymers of *d*-ribose and

[13] Berthelot, M. (1858), *Ann. Chim. Phys.*, 54, 74.
[14] Musculus, M. (1872), *Bull. Soc. Chim.* (*France*), 18, 66.
[15] Manners, D. J. (1965), Mereer, G. A., and Rowe, J. J. M., *J. Chem. Soc.*, 2150.
[16] Hilgetsy, G., Schramm, G., Martini, A., and Teichmann, H. (1951), *Chem. Ber.*, 91, 1073; — (1961), *Angew. Chem.*, 73, 619; — (1962), 74, 53; — (1964), *Biochem. Biophys. Acta*, 80, 1.

d-fructose with molecular weights of 40 000 have been obtained. This is a considerable achievement and implies that starch-like material can be synthesised and its chemical configuration can be controlled by the appropriate adjustment of the thermodynamic equilibrium in the acid medium in which the reaction is carried out.

Other chemical methods for synthesising, if not starch, at least polysaccharides of various molecular weight and with structures rendering at least some of them edible, have been worked out. Polysaccharides can be formed if dry, crystalline *d*-glucose is allowed to react with hydrogen chloride gas.[17] Another polymer, a synthetic glucan, can be produced when maltose is treated in much the same way. But the process developed by O'Colla and Lee[18] is in many ways more interesting still. These workers prepared several oligosaccharides and polysaccharides by heating *d*-glucose in the presence ion-exchange resins. By the choice of an appropriate resin—'Amberlite IR-120' was found to give good results—a significant control over the progress of polymerisation, and hence over the nature and yield of the polysaccharide produced, could be obtained.

A further series of synthetic oligosaccharides and starches has been synthesised by heating mixtures of dry sugars in a vacuum.[19] Considerable progress has been made in this method of polymerisation.[20] A high degree of polymerisation was achieved and various products obtained when *d*-glucose was heated by infra-red rays until it melted in such a way that oxygen was entirely absent, and any water derived from the reaction removed.[21] All sorts of structural forms were found in the synthetic glucans produced. There were several types of glucose links and many of the polymerised chains were highly branched.[22]

These, and other methods for synthesising sugar polymers, have been reviewed in considerable detail by Goldstein and Hullar.[23] But, having done so, they reached the conclusion that 'although a synthetic polyglucose is currently in use as an adhesive, the probability of utilisation of most of the synthetic polysaccharides for industrial

[17] Ricketts, C. R. (1954), *J. Chem. Soc.*, 4031.

[18] O'Colla, P. S. (1956), and Lee, E., *Chem. Ind.*, 522; — (1964), *J. Chem. Soc.*, 2351.

[19] Pictet, A. (1924), and Egan, M. M., *Helv. chim. Acta*, 7, 295.

[20] Pascu, E. (1950), and Mora, P. T., *J. Am. Chem. Soc.*, 72, 1045.

[21] Mora, P. T. (1958), and Wood, J. W., *J. Am. Chem. Soc.*, 80, 685.

[22] Dutten, G. G. S. (1964), and Unrau, A. M., *Can. J. Chem.*, 42, 2048.

[23] Goldstein, I. J. (1966), and Hullar, T. L., *Adv. Carbohydrate Chem.*, 21, 431.

purposes seems uncertain'. And they went on, 'The availability of vast stores of starch and other natural gums and of the knowledge and technology necessary to modify these substances into useful products, make it unlikely that much commercial interest will become focused on synthetic polysaccharides.' Indeed, so far the principal use to which artificial carbohydrates has been put has been for further research into the chemical mechanisms involved in their making or into their effect when employed to induce particular pharmacological ends. One minor use for glucans of known molecular weight has been as plasma substitutes in blood transfusion.

Hunger is always present in the world. The Bible is full of references to the time when the people cried out for bread. Today there are particular food needs for those at the extreme poles of our social system—among the very poor who have nothing and among those of all others the most affluent, namely the astronauts, for whom, in literal truth, money is no object. At present, and with reference to the ordinary world, there is no arguing with Goldstein and Hullar. Starch, or its equivalent synthesised from sugars, themselves made from formaldehyde—even assuming that such synthesis can be worked up into an industrial process—will always be more expensive under normal circumstances than natural starch from cereals or roots. But circumstances may change. It is not long since pineapples and other tropical fruits were exotic rarities in the countries of the northern hemisphere. Then it could have been said, as Goldstein and Hullar say now, that it was unlikely that much commercial interest would ever become focused on them as commonplace food-stuffs. Further, remarkable as are the possibilities of making formaldehyde from coal or oil, sugars from formaldehyde, and dextrins and starches from sugars, there is one more possibility that is more remarkable still.

Sugars and starches are plentiful in nature and in consequence cheap because during the course of biological evolution the mechanism of photosynthesis was evolved. The central feature of this process, upon which the whole of higher life-forms depend, is the multi-stage process by which the energy of sunlight combines the carbon dioxide of the atmosphere with water, wrests off the oxygen to return it to the air, and leaves the re-energised carbon and hydrogen in the form of sugars. Anyone who has sucked a grass stalk has had a taste of the sugar synthesised in the green leaf. In calculating the maximum theoretical yields of dry matter from any particular crop, horticulturalists calculate the proportion of the sun's energy intercepted by its leaves, the proportion of carbon dioxide in the atmosphere (or in the air within a greenhouse) and then assess the percentage efficiency with which the crop is converting the carbon dioxide into sugars and thence into plant substance. Considering the subtlety of the chemical conversion, the efficiency of leaf chlorophyll as a means of bringing about photosynthesis is high.

Between 1921 and 1928, a distinguished chemist, E.C.C. Baly, published a series of papers in which he described how by bubbling carbon dioxide gas through water and irradiating the solution with ultra-violet light he had been able to produce formaldehyde.[24] The implications of these experiments, taken at their face value, are profound. By means of comparatively simple techniques, Baly and his colleagues were claiming to be able to convert the physical energy of ultra-violet irradiation into the chemical energy contained in the formaldehyde molecule, which, as we have already seen, can be condensed into sugars—and to do this without the intermediacy of chlorophyll. If this should be proved true, then indeed could food be synthesised in a factory. Two curious puzzles, however, hang over Baly's discovery. For seven years he, a reputable and distinguished scholar, together with several colleagues, published the results of his researches in substantial journals, including that of the Royal Society. Then the reports stopped. It is said—I do not know with how much truth—that he then became unable to repeat his earlier work because, it was thought, of some change in the glass through which the irradiation was done. This is the first puzzle. But the second, to my mind, is why, now that almost half a century has passed and that greatly improved equipment is available, another team of investigators does not re-open the studies initiated by Baly. The prize for success could be great.

In spite of the pessimistic views of Goldstein and Hullar that no matter how efficient methods for synthesising sugars become or how readily means are found to polymerise sugar units to make dextrins and starches, such material will never be able to compete in price with starch from plant sources, there still may be something more to say. It was shown in Chapter 2 that protein produced by growing

[24] Baly, E. C. C. *et al.* (1921), *J. Chem. Soc.*, 119, 1025; — (1927), *Proc. Roy. Soc. Lond., A*, 116, 197, 212; — (1928), *Science*, 68, 364.

yeast on petroleum is, up till the present date at least, more expensive than soya protein or protein from groundnuts. There is, nevertheless, reason to continue work on its production. To start with, the process might be capable of improvement, with a consequent reduction in the price of the final product. Alternatively, a petroleum-protein plant might be conveniently set up in a part of the world where world prices cannot appropriately be applied and where the protein would, therefore, be to the profit of local consumers. Both these arguments could be applied to a synthetic sugar factory.

Another driving force behind attempts to propagate yeast on petroleum is that it constitutes a way of disposing of undesirable sludge fractions that accrue in refineries. If these effluents have to be got rid of in any event, it is clearly advantageous to make something useful, like yeast protein, out of them rather than dispose of them by some other means which might, in fact, involve as much expense as any charge arising from marketing protein at less than its full cost of manufacture. Again, the same argument could apply to synthetic sugar. If sugar and other carbohydrates come to be synthesised on a large scale, it seems likely—taking the pessimistic view that Baly's direct synthesis from carbon dioxide cannot be repeated—that substantial tonnages of formaldehyde from petroleum, natural gas or coal would be needed as raw material and would, consequently, be one charge on the cost of manufacture. It is, therefore, interesting to note that some of the samples of industrial effluent flowing into the river Thames in 1967[25] contained 4620 mg of formaldehyde per litre, and constituted a serious pollution hazard. Here then is an opportunity for industries producing such effluents to avoid polluting the environment and at the same time make a contribution to the supply of food.

[25] Greater London Council (1967), *Ann. Rep., Sci. Adviser.*

CHAPTER 5

SYNTHETIC VITAMINS

There are several good reasons why vitamins were the first food components to be manufactured as synthetic compounds and successfully marketed as fine chemicals. First, vitamins, by definition, are substances which are present in food in small amounts, yet these small amounts are of critical physiological significance. In other words, vitamins are valuable in relation to the quantities in which they occur. On the other hand, whereas the main components of food, particularly starch and protein, are polymers of large molecular weight, the vitamins, although of unusual chemical configuration at the time of their discovery, are substances of comparatively small molecular size. But the factor which led most directly to their being synthesised early on a commercial scale was that as soon as their existence was discovered, investigation was immediately started to find out their chemical nature. For example, when it was found that rice polishings were a rich source of what was then called "vitamin B", researches were undertaken to extract the vitamin from the rice polishings and obtain it in a more concentrated form. It was got into solution and adsorbed on to fuller's earth; then it was eluted from the fuller's earth and the elution liquor purified in a series of steps. At each stage. the consecutive concentrates were tested on rats or pigeons which had previously been maintained on a diet known to be lacking in 'vitamin B'. Later, the work was very much expedited when B. C. P. Jansen[1] observed that the active fractions always fluoresced in ultra-violet light after having been treated with potassium ferricyanide. Before long, pure crystalline 'vitamin B' was obtained. This was the first major step in the discovery of the mechanism of its physiological activity.

The next step in the research was the prolonged and laborious investigation undertaken in several laboratories to establish its chemical identity. Yet even when this was established by the accepted methods of organic chemical analysis, the evidence for the structure

[1] Jansen, B. C. P. (1936), *Rec. trav. chim.*, 55, 1046.

proposed was only circumstantial. Proof of the proposed configuration of a chemical compound, up till that time unknown to science, could only be confirmed by synthesising the molecule and then examining the synthetic compound obtained to see whether or not it possessed the same effectiveness to remedy the symptoms of vitamin deficiency in rats or pigeons—and finally in man—as the natural compound. When this was done, not only was confirmation of the correctness of a scientific deduction achieved, but at the same time a ready-made process was available by which the vitamin could be prepared by chemical synthesis.

This kind of approach, with variations for each individual compound, was quickly productive and new knowledge rapidly accrued. Vitamin C (ascorbic acid) was synthesised in 1933 by Reichstein[2] and by Haworth and his collaborators,[3] riboflavine, once called vitamin B_2, was synthesised in 1935 by Kuhn[4] and Karrer,[5] and in the same year Windaus and his colleagues[6] synthesised *d*-7-dehydrocholesterol, the natural substance which, on being irradiated, becomes vitamin D_3. Before this, Steenbock[7] had discovered in 1924 that vitamin D activity could be induced in what was soon recognised to be ergosterol, a substance which had been originally isolated by Braconnot in 1811. It can be seen from all this and much more—vitamin A was synthesised in 1937, vitamin B_6 (pyridoxin) in 1939 and nicotinic acid, only recognised as the pellagra-preventative part of the vitamin B_2 complex in 1937, was originally prepared as a pure substance in 1870—that the artificial manufacture of food components, whether or not it may constitute a major contribution to the food of the future, has very respectable antecedents in the past.

The Scottish naval surgeon James Lind recognised that oranges contained an active therapeutic agent capable of curing scurvy in sailors as long ago as 1753,[8] when he carried out an experimental trial which was a model of its kind on board the *Salisbury* at sea. But it

[2] Reichstein, T. (1933), *Nature, Lond.*, 132, 280.

[3] Ault, R. G. (1933), Baird, D. K., Carrington, H. C., Haworth, W. N., Herbert, R. W., Hirst, E. Z., Percival, E. G. V., Smith, F., and Stacey, M., *J. Chem. Soc.*, 1419.

[4] Kuhn, R. (1935), and Weygand, F., *Ber.*, 68, 1001.

[5] Karrer, P. (1935), Becker, B., Benz, F., Frei, P., Salomon, H., and Schopp, K., *Helv. chim. Acta*, 18, 1435.

[6] Windaus, A. (1935), Lettre, H., and Schenck, F., *Ann.*, 520, 98.

[7] Steenbock, H. (1924), *Science*, 60, 224.

[8] Lind, J. (1753, rep. Edinburgh U.P., 1953), *A treatise of the scurvy.*

was only in 1928 that Albert Szent-Gyorgyi,[9] after numerous other workers had carried out a long and frustrating series of trials, eventually succeeded in isolating the pure substance in crystalline form from adrenal glands, from oranges and from cabbage. And it was 1933 before its chemical structure was established.[10] Although at the time it was discovered it was an unusual structure, it is a comparatively simple molecule made up of the 6-carbon-atom chain shown below:

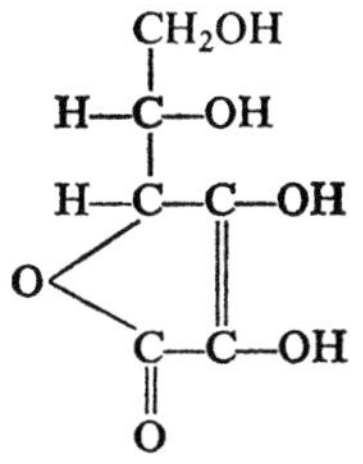

```
          CH2OH
           |
       H—C—OH
           |
       H—C—C—OH
         /     ||
     O<        ||
         \     ||
           C—C—OH
           ||
           O
```

Ascorbic acid (*vitamin C*)

Once the chemical configuration of ascorbic acid had been established, it was not long before the expertise and virtuosity of the organic chemists who tackled the problem discovered several methods by which it could be made artificially by chemical synthesis. One method uses the sugar, *l*-sorbose, as a starting material. *l*-Sorbose occurs in nature and can be isolated from berries of the mountain ash (*Sorbus aucuparia*). A more practical source of raw material is glucose which can be converted into the alcohol, *d*-sorbitol, by catalytic hydrogenation. The *d*-sorbitol can then be transformed into *l*-sorbose by means of a fermentation process using acetic acid bacteria.[11] Several ways have been found for oxidising *l*-sorbose into 2-ketogulonic acid, the next stage in the synthesis. For example, Haworth[12] achieved this step by oxidation using nitric acid. The 2-ketogulonic acid can then be 'lactonised' to produce ascorbic acid itself by several different methods.[13]

[9] Szent-Gyorgyi, A. (1928), *Biochem. J.*, 22, 1387.

[10] Herbert, R. W. (1933), Hirst, E. Z., Percival, E. G. V., Reynolds, R. J. W., and Smith, F., *J. Soc. Chem. Ind.*, 52, 221, 481.

[11] Bertrand, G. (1896), *Bull Soc. Chim.*, 15, 627; Wells, P. A. (1939), Lockwood, L. B., Stubbs, J. J., Porges, N., and Gastrock, E. A., *Ind., Eng. Chem.*, 31, 1425.

[12] Haworth, W. N. (1934), *Nature, Lond.*, 134, 724; British Patent 443,901.

[13] Reichstein, T. (1934), and Grüssner, *Helv. chim. Acta*, 17, 311.

An alternative method of synthesising ascorbic acid is to start from a 5-carbon-atom base instead of from a 6-carbon-atom one. The most convenient such starting compound is *l*-xylose, which occurs in nature as part of the hemicellulose which is a component of such things as corncobs, sawdust from various kinds of softwood and coconut shells. Xylose can also be prepared by degrading the 6-carbon-atom molecule of glucose. This can also be done with *d*-sorbitol. To synthesise ascorbic acid from xylose,[14] the xylose is first oxidised to yield *l*-xylosone and subsequently treated with hydrogen cyanide, by which the appropriate 6-carbon-membered structure of ascorbic acid is obtained.

The structures of the two substances *l*-sorbose and *l*-xylose, are shown below:

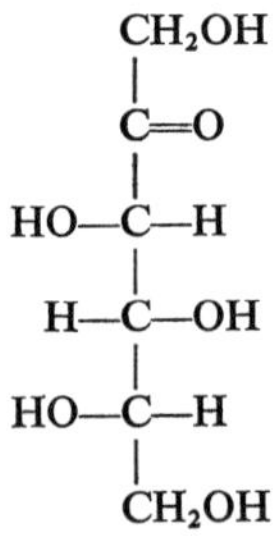

$$\begin{array}{c} \mathrm{CH_2OH} \\ | \\ \mathrm{C{=}O} \\ | \\ \mathrm{HO{-}C{-}H} \\ | \\ \mathrm{H{-}C{-}OH} \\ | \\ \mathrm{HO{-}C{-}H} \\ | \\ \mathrm{CH_2OH} \end{array} \qquad \begin{array}{c} \mathrm{CHO} \\ | \\ \mathrm{HO{-}C{-}H} \\ | \\ \mathrm{H{-}C{-}OH} \\ | \\ \mathrm{HO{-}C{-}H} \\ | \\ \mathrm{CH_2OH} \end{array}$$

l-Sorbose 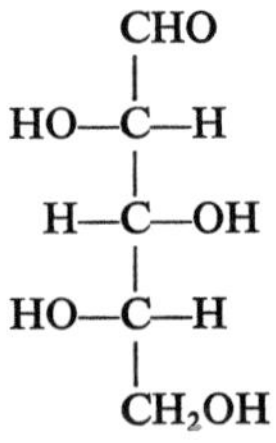 *l-Xylose*

Several other alternative routes to the synthesis of the vitamin have been proposed,[15] but the most usual starting point in commercial synthesis has been *l*-sorbose. Before long, very substantial quantities of ascorbic acid were being manufactured and the purity of the synthetic vitamin was greater than 99%. It quickly became very much cheaper to prepare concentrates of the vitamin from synthetic material rather than from natural sources such as oranges or blackcurrants. By 1956, the price was 12 cents per g; six years later in 1962 it was 6 cents per g.[16] For a consumer to obtain a gram of

[14] Reichstein, T. (1933), *Nature, Lond.*, 132, 280; Reichstein, T. (1934), Grussner, A., and Oppenauor, R., *Helv. chim. Acta*, 17, 510; Ault, R. G., *et al.* (1933), *J. Chem. Soc.*, 1419.

[15] Helferich, B. (1937), and Peters, O., *Ber.*, 70, 465; Michael, F. (1940), and Haarhoff, H., *Ann.* 545, 28.

[16] Fox, S. W. (1963), *Food Tech.*, 17, 4.

ascorbic acid from a natural source, would require the consumption of about 1 lb of blackcurrants, 4 lb or oranges or 60 lb of apples.

The comparatively low cost of synthetic ascorbic acid has led to its wide use in medicine and as a nutritional supplement. It is added as routine to numerous formulae for infant feeding. It is added to fruit-based soft drinks. But ascorbic acid is not only well established as a synthetic food component, it is also used on a large scale as an 'additive', not on account of its nutritional significance, but purely for the technological effects which can be obtained from its chemical properties. For example, it is used in baking as a bread 'improver' to benefit the appearance and crumb-structure of the loaf. It is incorporated in sausages and in other meat products to improve their appearance. It imparts a brighter red to raw meat if it is rubbed on to it, although some doubts have been expressed as to whether this might not be construed as a means of passing off meat as being fresher than it actually is. It is used in the treatment of wine, employed also in brewing, and used as an antioxidant in the manufacture of preserved vegetables. None of these uses contributes any nutritional benefit either to bread, wine, meat or vegetables, since the vitamin becomes oxidised in achieving its technological result. Its further use as a reducing agent in developing photographic negatives is also palpably non-nutritional.[17]

It is clear from all this that synthetic ascorbic acid has established a firm place as a significant—indeed a substantial—market source of the total vitamin C requirements of industrial communities. It is particularly valuable for pharmaceutical uses. Its low cost has led to its employment in enormous doses of more than 1 g per day as a prophylactic against colds, although there is no convincing scientific evidence of its effectiveness for this purpose; the daily nutritional requirement is only between 0·03 and 0·07 g. Where it is used as a technological adjunct, although it contributes nothing of nutritional value, its use at least raises no possibility of toxicological danger, which could arise when substances alien to biological functioning are employed. Synthetic ascorbic acid must, therefore, be accepted as an item of fine-chemical manufacture of considerable economic importance.

[17] Wiss. Ver. Deutch Ges f Enah, (1965). *11th Symposium on Ascorbic Acid*, Steinhopff, Darmstadt.

Synthetic ascorbic acid possesses identical antiscorbutic effectiveness to the natural vitamin in fruit as measured in laboratory tests on guinea-pigs or in medical practice. The comparison of concentrated orange juice or some other natural source with a preparation of the synthetic substance on a basis of equal ascorbic acid concentration, however, raises certain problems of interpretation. On the one hand, ascorbic acid is not the only nutritionally active component of orange or grapefruit juice, and deficiency states are commonly found, when they occur, to be due to an impoverished diet lacking in more than one component. It is for this reason that preparations of orange juice or some other natural source of ascorbic acid are often preferred to the synthetic vitamin for infant feeding.

Although evidence of the existence of B vitamins does not date back as far as Lind's studies of scurvy showing the curative effect of what was later called vitamin C in oranges, it does go back to 1884. It was then that Takaki[18] prevented beri-beri in the Japanese navy by changing the sailors' diet. By 1926, Jansen and Donath[19] succeeded in isolating pure vitamin B_1, subsequently called thiamine, from rice polishings. In 1936, Williams[20] and Greive[21] elucidated its chemical structure, and in the same year Williams and Cline[22] and Andersag and Westphal[23] both succeeded in working out a method by means of which it could be synthesised. But while all this was going on, experimental studies on rats and pigeons, in which yeast and rice polishings and concentrates derived from them were used as sources of the vitamin, quickly showed that what had first been taken to be a single substance and called 'vitamin B' was in fact a group of substances. Until this was known, and the detailed chemistry of each member understood and its particular physiological significance determined, a natural source of 'vitamin B' was a better supplement to a nutritionally deficient diet than the single pure crystalline thiamine (designated vitamin B_1) alone. The chemical structure of thiamine is shown opposite.

The synthesis of thiamine involved first of all the synthesis of the

[18] Takaki, K. (1885), *Sci-i-Kai Med. J.*, Aug.; *ibid.* (1886), Apr.; *ibid.* (1887), 6, 73.

[19] Jansen, B. C. P. (1926), and Donath, W. F., *Med. Dienst Volks. Ned. Ind.*, *Pt 1*, 186.

[20] Williams, R. R. (1936), *J. Am. Chem. Soc.*, 58, 1063.

[21] Greive, R. Z. (1936), *Physiol. Chem.*, 242, 89.

[22] Williams, R. R. (1936), and Cline, J. K., *J. Am. Chem. Soc.*, 58, 1504.

[23] Andersag, H. (1937), and Westphal, K., *Ber.*, 70, 2035.

```
 ┌                                                          ┐⊕
         H              CH2             CH3
       N⁄C⸌C ──────────⁄   ⸌N───C⁄
       ‖     |                ‖     ‖                         Cl⊖
 CH3─C⸌N⸗C─NH2·HCl       CH    C─CH2·CH2OH
                                 ⸌S⁄
 └                                                          ┘
```

Thiamine (vitamin B_1)

first ring structure, the so-called 'pyrimidine moiety' of the molecule, then the preparation of the second sulphur-containing thiazole ring and, finally, the linking of the two together to produce the complete molecule. These steps were quite quickly achieved by several groups of expert organic chemists working in Great Britain, Germany, the United States and elsewhere. Several different chemical routes were followed, some of which were patented.[24] It was not long before synthetic thiamine became readily available.

The first use of the synthetic material was in pharmacy, and particularly for medical treatment. Although tablets containing thiamine formed a convenient vehicle by which the vitamin could be administered, there was good reason for doctors to prefer tablets of dried yeast as a source, not only of thiamine but also of other components of the vitamin B complex. But for administration by injection, for patients seriously ill with beri-beri or suffering from alcoholic neuritis, ampoules containing a solution of pure synthetic thiamine were clearly preferable. Very soon, however, the availability of synthetic thiamine in quantity made it feasible to use it as a direct source of the nutrient in bulk food supplies. In 1939, it was planned to provide a wartime loaf in Great Britain made of white flour enriched with supplementary synthetic thiamine. It was argued that this would provide a popular food-stuff for the people. By using white bread for human food, a source of milling by-products would become available to feed pigs and thus provide bacon, while demanding only a very small importation of specialist chemical intermediaries for the factory carrying out the synthesis. It was only the intensification of the war in 1940 that made it necessary for the scheme to be postponed and for the citizens themselves, instead of the pigs, to eat a proportion of the milling by-products in the form of bread of raised extraction

[24] German Patents 667,990; 670,635; 671,787; 664,789; 669,187; 676,980; U.S. Patent 2,160,867; 2,184,720; 2,127,446; French Patent 831,110; 816,432.

rate. By the 1950s, however, synthetic thiamine had been accepted as a normal ingredient of standard food-stuffs. In 1956, on the recommendation of a Government Panel on the Composition and Nutritional Value of Flour, the British public-health authorities laid down that white flour must contain 0·24 mg of thiamine per 100 g. What was not naturally present in the flour was added as the synthetic vitamin. In the Philippines and elsewhere in Asia, rice was similarly artificially enriched with thiamine and other vitamins manufactured for the purpose. Between 1956 and 1962, the price of synthetic thiamine fell from 60 cents to 17 cents per g.

The next component of the vitamin B complex, riboflavin, was synthesised in 1935 by two groups of investigators. One was in Germany, headed by Kuhn;[25] the other, in Switzerland[26] under the direction of Karrer used xylene (derived from petroleum) and the sugar, D-ribose, as starting materials. The chemical structure of the vitamin is shown below:

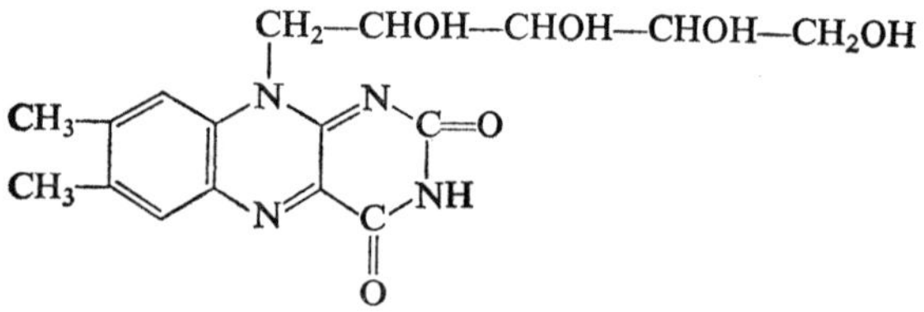

CH_2—CHOH—CHOH—CHOH—CH_2OH
N N C=O
CH_3
CH_3
N C NH
O

Riboflavin

The chemical synthesis of so comparatively elaborate a molecule as that of riboflavin is quite a complicated operation. It followed, therefore, that the economics of manufacture by chemical synthesis had to be considered in relation to the cost of preparing it from natural sources or by biological means. The degree of purity required had also to be taken into account. For animal feeding, for example, riboflavin concentrates prepared from yeast, whey or, more particularly, from fermentation, were cheaper and more convenient than direct chemical manufacture. As long as the bulk organic chemicals, acetone and butanol, were manufactured—as they were for many years—by a fermentation process, riboflavin could be cheaply manufactured from the bacterial residues remaining from the acetone–

[25] Kuhn, R. (1935), and Weygand, F., *Ber.*, 68, 1001.
[26] Karrer, P., *et al.* (1935), *Helv. chim. Acta*, 18, 1435.

butanol process. Even when these substances were later produced from petroleum, a fermentation process using *Eremothecium ashbii* was developed by which riboflavin could be prepared at a comparatively cheap price. For human therapy, however, where material of a high degree of purity is required, direct chemical synthesis of the vitamin has a part to play. By 1954, the production of crystalline riboflavin in the United States alone had reached 266 000 lb and between 1956 and 1962 the price fell from 65 cents to 33 cents per g.

In the 1930s, as one after another of the various B vitamins were isolated, purified diets could be made up and fed to experimental animals, and, as each time the added fractions were found insufficient to make up a fully satisfactory diet, first the existence and then the identity of the next vitamin B fraction was elucidated. Vitamin B_6, recognised first as an anti-dermatitis factor in rat diets,[27] but later found to be different from the pellagra-preventative vitamin,[28] was identified and synthesised in 1939. Its chemical structure was found to be that shown below:

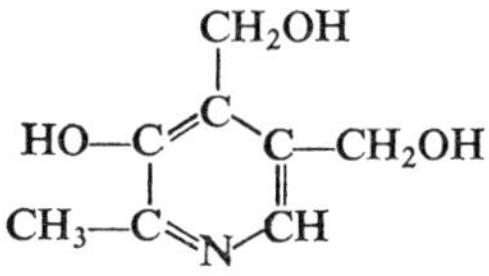

```
            CH2OH
            |
            C
         //   \
HO—C           C—CH2OH
     |            ||
CH3—C          CH
         \\   /
            N
```

Pyridoxin (vitamin B_6)

Two methods of synthesising pyridoxin have been developed. In the first,[29] the comparatively simple chemical intermediates cyanoacetamide, NH_2—CO—CH_2—CN, is condensed with ethoxyacetylacetone, CH_3—CO—CH_2—CO—$CH_2OC_2H_5$, to form a ring structure from which the exact configuration of the vitamin is derived. In the second,[30] the ring structure, potassium phthalimide is condensed with ethyl 1-bromopropionic acid ester to produce the initial starting structure from which the molecule of pyridoxin is built up.

Although a good deal is known about the function of pyridoxin

[27] Goldberger, J. (1926), and Lillie, R. D., *U.S. Pub. Health Serv. Rep.*, 41, 1025.
[28] Gyorgyi, P. (1934), *Nature, Lond.*, 133, 498.
[29] Harris, S. A. (1939), and Folkers, K., *J. Am. Chem. Soc.*, 61, 1245; Morii, S. (1939), and Makino, K., *Enzymologia*, 7, 385.
[30] Kuhn, R. (1939), Westphal, K., Wendt, G., and Westphal, O., *Naturwissenschaften*, 27, 469; Itiba, A. (1939), and Miti, K., *Sci. Pap. Inst. Phys. Chem. Res., Tokyo*, 36, 173.

in human metabolism—it plays a part in the utilisation of amino-acids and in the metabolism of nervous tissue—and deficiency has been found to occur in infants under abnormal circumstances, such deficiency is very rare. The dietary requirement of normal infants is recognised to be very minute. No one has been able to establish what the needs of pyridoxin are in the diets of adults, or even whether adults can ever suffer from dietary deficiency at all. Hence, although the vitamin can be synthesised and is indeed manufactured as a comparatively rare chemical, the market for it is small.

Nicotinic acid, also called niacin, and its amide, nicotinamide, are, in certain respects, different from other members of the vitamin B complex. Populations who eat a niacin-deficient diet, and particularly those who consume a deficient diet in which maize is a major ingredient, suffer from a serious disease from which many may die and more may be afflicted with some or all of the debilitating symptoms, often summarised under the headings, dermatitis, dysentery and dementia. At the same time, the chemical structure of nicotinic acid, identified in 1937 as the substance whose lack is the main cause of pellagra,[31] was first determined in 1870.[32] The chemical structure of nicotinic acid, nicotinamide and nicotine are shown below:

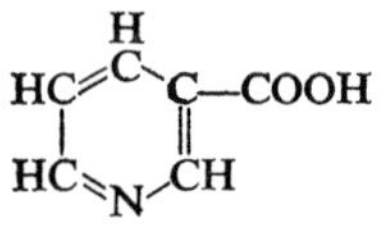

Nicotinic acid	*Nicotinamide*	*Nicotine*
H, HC=C—C—COOH, HC=N—CH (pyridine ring)	H, HC=C—C—CONH$_2$, HC=N—CH (pyridine ring)	H, HC=C—C—HC(—H_2C—CH_2—CH_2—N—CH_3), HC=N—CH (pyridine ring)

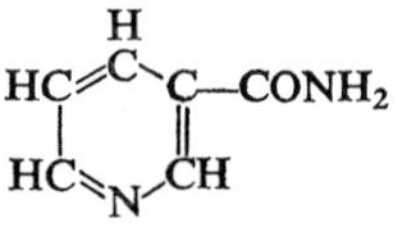

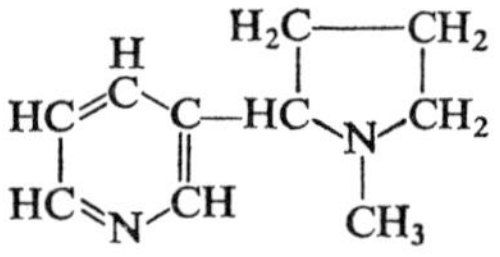

Huber, in 1870, first discovered nicotinic acid after having oxidised a preparation of nicotine from tobacco. The nicotinic acid produced readily forms crystallisable salts which can easily be recovered from the oxidation mixture and purified. It is interesting to note that one of the most convenient modern methods for the manufacture of nicotinic acid in quantity is to treat nicotine with fuming nitric acid, chromic acid or permanganate. This process can hardly be defined as synthesis: nor can it be claimed to add to any marked extent to the world's food production. A significant area of the world's fertile

[31] Fouts, P. J. (1937), Helmer, O. M., Lepkovsky, S., and Jukes, T. H., *Proc. Soc. Exp. Biol. Med.*, 37, 405.

[32] Huber, C. (1870), *Ber.*, 3, 849.

agricultural land, not only in the wealthy United States but in much poorer Turkey and India as well, is devoted to the culture of tobacco when it might be used for food. It is a bizarre twist to the age-old struggle to provide nourishment for the subsistence of mankind, to devote land, first of all, to the culture of a non-food drug of some recognised toxicity, and subsequently convert part of it by technological means into a 'synthetic' vitamin.

Nicotinic acid and nicotinamide can be seen to be comparatively simple chemical compounds. There are, therefore, several ways now known by which they can be synthesised from quite simple starting materials. One method is to start with pyridine derived from coal tar. The pyridine is first brominated, then converted into 3-cyanopyridine by means of cuprous cyanide, and then saponified to convert it to nicotinic acid.[33]

Nicotinic acid can readily be converted into nicotinamide. One of the most direct methods is to treat it with gaseous ammonia at an elevated temperature.[34]

Large amounts of niacin are added to a variety of manufactured foods, sometimes to supply an adequate amount of the vitamin to a diet calculated to be deficient, sometimes (as when it is incorporated in a commodity such as corn flakes) with the general idea that it might conceivably be useful, regardless of whether or not evidence of shortage in the consumer's intake is available. By the early 1960s, the price of synthetic niacin was about 6 cents per g.

Yet one more member of the vitamin B complex, pantothenic acid, whose existence was first suggested in 1928[35] on evidence of experiments on pigeons, subsequently extended to rats, was identified in 1939[36] as the compound

$$HOCH_2\cdot\underset{\displaystyle CH_3}{\overset{\displaystyle CH_3}{\overset{|}{\underset{|}{C}}}}—CHOH\cdot CO\cdot NH\cdot CH_2\cdot CH_2\cdot COOH$$

[33] McElvain, S. M. (1941), and Goese, M. A., *J. Am. Chem. Soc.*, 63, 2283.

[34] Keimatou, S. (1933), Yokata, K., and Satoda, A., *J. Pharm. Soc. Japan*, 53, 994.

[35] Williams, R. R. (1928), and Waterman, R. E., *J. Biol. Chem.*, 78, 311.

[36] Williams, R. J. (1939), Weinstock, H. H., Rohrmann, E., Truesdail, J. H., Mitchell, H. K., and Meyer, C. E., *J. Am. Chem. Soc.*, 61, 454.

Within a year it had been synthesised.[37] The starting material was the amino-acid β-alanine, itself synthesised from acetaldehyde, sodium cyanide and ammonium chloride. By the mid-1950s the production in the United States of pantothenic acid was 286 000 lb per year, by the mid-1960s the price had dropped to 2 cents per g—and the remarkable thing is that it is naturally present in almost every kind of available food-stuff, and no one has so far been able to diagnose with any degree of certainty that any human being has ever suffered from going short of it.

Two further vitamins which are grouped with the so-called vitamin B complex are worthy of comment in an account of the chemical synthesis of these factors. They are folic acid and vitamin B_{12}.

During the 1930s, Dr Lucy Wills, who was studying nutritional anaemias in pregnant women in Bombay, observed that a type of anaemia that she was finding, which did not respond to any vitamin then known nor to liver extract, was strikingly benefited by treatment with an extract prepared from yeast.[38] In due course, this factor in yeast was found to be the same as a substance derived from spinach—hence the name 'folic acid'[39]—which stimulated the growth of the micro-organism, *Lactobacillus casei*. By 1945, this substance had been isolated in crystalline form and its chemical configuration established.[40] Its structure is shown below:

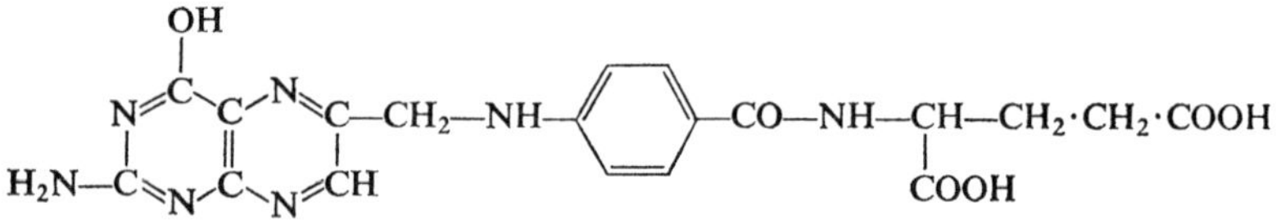

```
            OH
            |
        N=C   N
      N=   C-C  =C—CH2—NH—⟨benzene⟩—CO—NH—CH—CH2·CH2·COOH
      |      ||   |                          |
  H2N—C=N-C-N=CH                            COOH
```

Folic acid (pteroylglutamic acid, PGA)

Although the chemistry of this vitamin is by no means simple, its synthesis was not long delayed. The double-ring structure, pteroic acid, belongs to a group of compounds, the pteridines, originally

[37] Williams, R. J. (1940), Mitchell, H. K., Weinstock, H. H., and Snell, E. E., *J. Am. Chem. Soc.*, 62, 1784.

[38] Wills, L. (1933), *Lancet*, (*i*), 1283.

[39] Mitchell, H. K. (1941), Snell, E. E., and Williams, R. J., *J. Am. Chem. Soc.*, 63, 2284.

[40] Stokstad, E. L. R. (1946), Hutchings, B. L., Mowat, J. M., Boothe, J. H., Waller, C. W., Augier, R. B., Lamb, J., and Subba Row, J., *Ann. N.Y. Acad. Sci.*, 48, 269, 273.

discovered in the wings of butterflies. The constitution of pteridine was worked out in 1940. With the identity of folic acid discovered, intense activity was concentrated on its production, particularly by the powerful pharmaceutical firms in the United States. Parke Davis & Co. and Merck were involved, but it was Lederle Laboratories who achieved the greatest success in establishing the structure of the vitamin. To attain the final successful goal, however, involved them in expenses for research of more than $500 000.[41] Several similar compounds are now known to occur in nature. Pteroyltriglutamate and pteroylheptaglutamate contain three and seven glutamic acid residues, respectively, and folinic acid has an additional formyl group, —COOH, attached to the PGA molecule. The pure synthetic PGA, however, is physiologically active and is used in clinical and in veterinary medicine.

But although folic acid is concerned with the prevention of so-called megaloblastic anaemia, it is ineffective in the treatment of pernicious anaemia. For this, vitamin B_{12}, originally isolated from liver,[42] is required. This substance is active in very low concentration. Its structure is highly complex and was only elucidated after prolonged and highly gifted scientific research.[43] The structure of vitamin B_{12} is shown overleaf.

It is hardly surprising that a molecule of this complexity has defied synthesis. On the other hand, it *is* perhaps surprising that purified vitamin B_{12} can readily be produced commercially as a comparatively inexpensive by-product of the cultivation of *Streptomyces griseus* used in the preparation of the antibiotic, streptomycin.

So far, we have only discussed the water-soluble vitamins and their synthesis. But similar success has also been achieved with fat-soluble vitamins, and substantial amounts of the more important of these are currently synthesised for enriching food-stuffs and for use in clinical and veterinary medicine.

[41] Paterson, W. H. (1954), *Nutr. Rev.*, 12, 227.

[42] Smith, E. L. (1948), and Parker, L. F. J., *Biochem. J.* 43, viii; Rickes, E. L. (1948), Brink, N. G., Koniuszy, F. R., Wood, T. R., and Folkers, K., *Science*, 107, 396.

[43] Hodgkin, D. C. (1955), Pickworth, J., Roberston, J. H., Trueblood, K. N., Prosen, R. J., and White, J. G., *Nature, Lond.*, 176, 325; Bonnett, R. (1955), Cannon, J. R., Johnson, A. W., Sutherland, J., and Todd, A. R., *Nature, Lond.*, 176, 328.

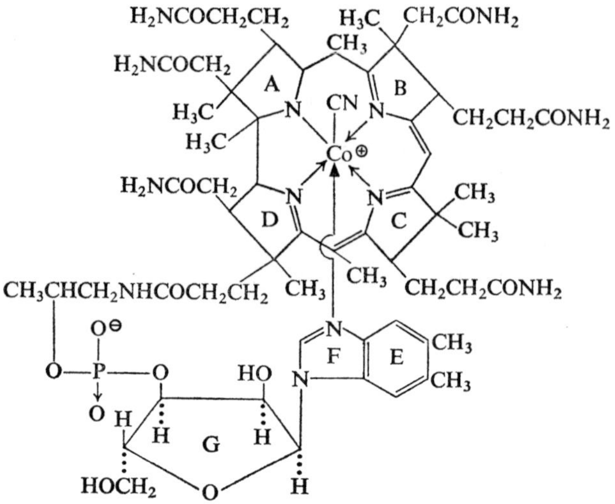

Vitamin B_{12} (cyanocobalamin)

Vitamin A, whose structure was determined by Karrer in Switzerland in 1931,[44] was synthesised in 1937.[45] This original synthesis was a complex operation involving several steps and, although a preparation of high activity was obtained, an entirely pure substance was not recovered. The chemical structure of vitamin A is shown below:

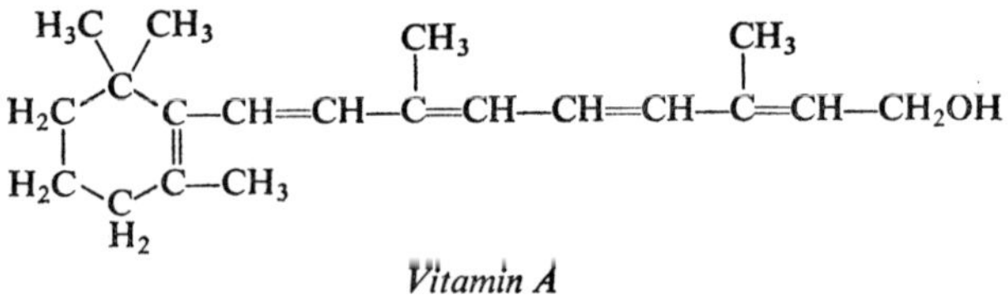

Vitamin A

The narrative of how vitamin A has been synthesised commercially represents a good example of some of the factors, both scientific and commercial, which are involved in food synthesis in general and in the synthesis of vitamins in particular.

When discussing vitamin C, it was pointed out that the synthetic material, being identical with that found in natural products, possessed

[44] Karrer, P. (1931), Morf, R., and Schopp, K., *Helv. chim. Acta*, 14, 1036, 1431.
[45] Kuhn, R. (1937), and Morris, C. J., *Ber.*, 70, 853.

the same physiological activity. While the same thing holds for vitamin A, it will only hold if the vitamin that is synthesised is in fact identical with what occurs in nature. For vitamin A, however, it was known, even before synthesis had been achieved, that more than one form of the vitamin existed in natural products. The variations in structure are not very great; nevertheless, it was found that the 'all-*trans*' form of vitamin A possesses a higher activity than the other forms. The different variations of molecular structure are shown diagrammatically in Figure 2.[46]

Enormous chemical ingenuity was devoted to the synthesis, not only of the all-*trans* form of the vitamin, but also to the 11-*cis* and the 11,13-di-*cis* forms as well as of vitamin A_2 containing an additional —CH_2OH group at the end of the chain. Methods of accomplishing these syntheses were elaborated by the firms of Hoffmann-La Roche in New Jersey and in Basle,[47] by the Badische Anilin Sodafabrik in Ludwigshafen,[48] and Distillation Products Industries in Rochester, New York.[49] Initially, the starting material for the synthesis of vitamin A was citral obtained from lemon grass oil. The total production quickly became very large, rising from 73·3 trillion USP (US production) units in 1952 to 106·9 trillion units in 1954. Obviously, this synthetic food-stuff was large-scale business and was subject to the influences affecting commerce in general, for example, fluctuations in the price of lemon grass, which in the United States varied from $1 to $4 per lb. It was not surprising to learn,[50] therefore, that in 1957, Hoffmann-La Roche had erected a multi-million dollar factory for manufacturing synthetic vitamin A by a process based on acetone and acetylene which rendered them independent of vagaries in the price of lemon grass oil. The synthetic process had thus become a strictly chemical one, and supplies of vitamin A were entirely independent, not only of such biological systems as the livers of calves or cod, but of lemon grass as well.

The very large amounts of vitamin A synthesised each year are employed primarily for what is their proper function, namely, the

[46] Isler, O. (1960), Ruegg, R., Schweitzer, V., and Wursch, J., *Vitamins and Hormones*, 18, 295.

[47] Kimel, W. (1957), Surmatis, J. D., Weber, J., Chase, G. O., Sax, N. W., and Ofner, A., *J. Org. Chem.*, 22, 1611; Saucy, G. (1959), Marbet, R., Lindler, H., and Isler, O., *Helv. chim. Acta*, 42, 1945.

[48] Pommer, H., and Sarnecki, W., Belgian Patent, 570,838.

[49] Humphlett, W. J., and Burness, D. M., U.S. Patent 2,676,990.

[50] Anon. (1957), *Chem. Trade J.*, 1056.

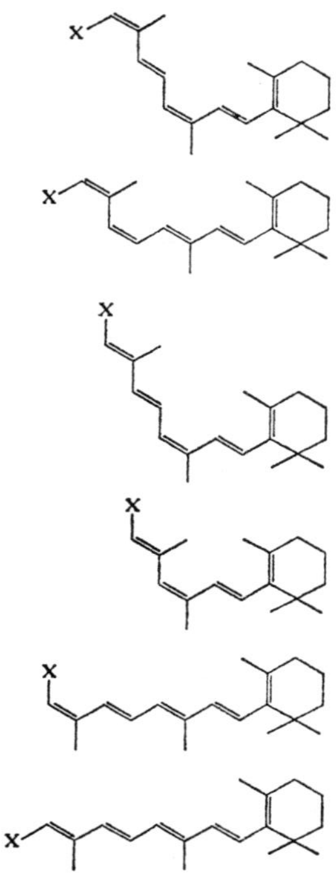

	Melting point (°C) of:		
	Alcohol	*Aldehyde*	*Acid*
all-*trans*	62–64	61–62	179–180
13-*cis* (neo *a*)	58–60	77	175–176
11-*cis* (neo *b*)	Oil	63·5–64·4	—
9-cis (iso *a*)	82–83	64	189–191
11,13-di-*cis* (neo *c*)	Oil	Oil	—
9,13-d-i*cis* (iso *b*)	58–59	49 85	135–136

Figure 2 **Different molecular structures of vitamin A**

enrichment of food-stuffs, and particularly margarine. Substantial quantities are also used in tablets and capsules and preparations for infant feeding. The situation for another fat-soluble vitamin, vitamin E (tocopherol) is rather different. Although the function of vitamin E in the tissues of the body is quite well understood, and although deficiency symptoms have been induced in rats and other experimental animals maintained on diets lacking in vitamin E, so far there is no unequivocal proof that vitamin-E deficiency ever occurs in man. Substantial amounts of synthetic vitamin are manufactured—in 1966, 507 000 lb valued at almost $8 000 000 were produced in the United States alone[51]—and some of it is used in pharmaceutical preparations. Because, owing to its chemical structure, it functions in nature as an antioxidant, the synthetic vitamin is mostly used, however, not directly to improve the nutritional value of foods to which it is added, but as a technological adjunct to inhibit their becoming oxidised and hence stale and rancid. So far as is known, no harm is caused from eating too much vitamin E.

Vitamin E, like vitamin A, is not a single substance but a family of compounds, each one varying from the other by small differences in composition or configuration.[52] Figure 3 shows the molecular structure upon which the group is based together with a list of some

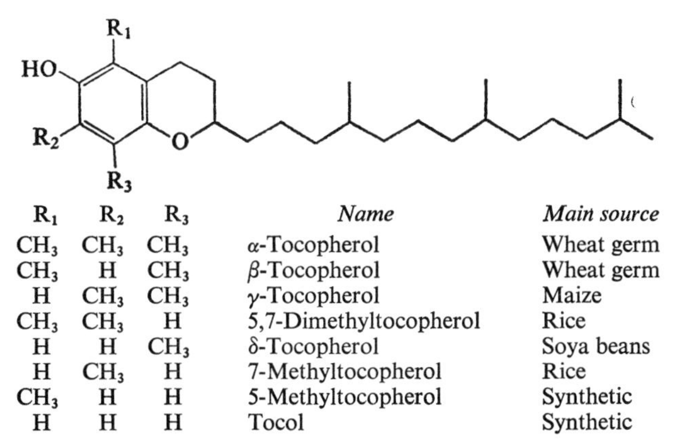

R_1	R_2	R_3	*Name*	*Main source*
CH_3	CH_3	CH_3	α-Tocopherol	Wheat germ
CH_3	H	CH_3	β-Tocopherol	Wheat germ
H	CH_3	CH_3	γ-Tocopherol	Maize
CH_3	CH_3	H	5,7-Dimethyltocopherol	Rice
H	H	CH_3	δ-Tocopherol	Soya beans
H	CH_3	H	7-Methyltocopherol	Rice
CH_3	H	H	5-Methyltocopherol	Synthetic
H	H	H	Tocol	Synthetic

Figure 3 The chemical configuration of some members of the vitamin E group (tocopherols)

[51] U.S.A. Synthetic Organic Chemists (1966), Table 13A, U.S. Dep. Comm., Bureau of the Census, Wash., D.C.

[52] Isler, O. (1962), Schudel, P., Mayer, H., Wursch, J., and Ruegy, R., *Vitamins and Hormones*, 20, 389.

of the variants. There is a further family of compounds in which the side chain has an unsaturated bond.

The synthesis of vitamin E-active compounds was first achieved in the laboratories of Hoffmann-La Roche in Switzerland, and the first total synthesis carried out by Karrer and his colleagues in 1938.[53] Since that time a series of improvements and modifications have been introduced, so that any variant of the vitamin can be manufactured on demand.[52]

Vitamin K, like vitamin E, is a substance which, so far as is known, is always present in adequate amount in the diet of normal people, no matter how inadequate in other respects that diet may be. Alternatively, it can be said that whatever amount of vitamin K an ordinary individual may require is synthesised for him by the micro-organisms naturally present in his gut. There are two circumstances, however, when a supply of vitamin K from some outside source may be needed. Vitamin K is a fat-soluble substance. Like all other such compounds, bile salts are required for its absorption. It is sometimes found that a day or two after a patient has had an operation for the relief of jaundice due to obstruction of the bile duct severe bleeding occurs. Vitamin K, which is known to be concerned with the proper clotting of the blood, is nowadays always provided by injection before such operations are performed. The second proved need for vitamin K is in the treatment of some diseases (e.g., sprue, idiopathic steatorrhoea, ulcerative colitis) in which the absorption of fat is disturbed and when bleeding sometimes occurs.

The existence of vitamin K was originally discovered by research workers in Denmark,[54] who found that a condition in chickens in which bleeding occurred could be prevented or cured by change in diet. The new vitamin was isolated in 1935[55] and its synthesis accomplished almost simultaneously in three American laboratories in 1939.[56] The active material is found to occur in several different chemical forms as shown in Figure 4.

[53] Karrer, P. (1938), Fritzsche, H., Ringier, B. H., Salomon, H., *Helv. chim. Acta*, 21, 520, 820.

[54] Dam, H. (1934), and Schonheyder, F., *Biochem. J.*, 28, 1355.

[55] Dam, H. (1935), *Nature, Lond.*, 135, 652.

[56] Almquist, H. J. (1939), and Klose, A. A., *J. Amer. Chem. Soc.*, 61, 2557; Binkley, S. B. (1939), Cheney, L. C., Holcolm, W. F., McKee, R. W., Thayer, S. A., MacCorquodale, D. W., and Doisy, E. A., *ibid.*, 61, 2558; Fieser, L. F. (1939), *ibid.*, 61, 2559.

O, CH_3, R, O

$$\text{Vitamin } K_1\text{: } R = -CH_2\cdot CH{=}\overset{CH_3}{\underset{}{C}}\cdot CH_2[CH_2\cdot CH_2\cdot \overset{CH_3}{CH}\cdot CH_2]_2CH_2\cdot CH_2\cdot \overset{CH_3}{CH}\cdot CH_3$$

$$\text{Vitamin } K_2\text{: } R = -CH_2[CH{=}\overset{CH_3}{C}\cdot CH_2\cdot CH_2]_5CH{=}\overset{CH_3}{C}\cdot CH_3$$

Figure 4 Two forms of vitamin K

Significant amounts of vitamin K (in the K_1 form) are synthesised commercially. The amount produced in the United States in 1966 was 156 000 lb.[51]

The history of the manufacture of synthetic vitamin D provides several different lessons showing the influence of food synthesis for good and for ill. In some respects these lessons differ from those inherent in the synthesis of the vitamins that have already been described. In the absence of vitamin D, the disease of rickets occurs. Cod-liver oil was used as a traditional folk remedy in Scotland as long ago as the eighteenth century, and the French physician Trousseau used it as a specific remedy in the treatment of rickets about a century ago. But it was Mellanby[57] who, in 1918, first clearly showed in experiments with puppies that rickets was a nutritional disease which responded to a fat-soluble vitamin present in cod-liver oil. It was on the basis of this research that Chick and her colleagues carried out their classical work on children suffering from rickets in Vienna after the First World War, which led to the virtual extinction of the disease as the cause of crippling and disfigurement which it had up till then been.[58] Then, following the discovery by Steenbock in America that foods could be made curative of rickets (antirachitic) by being irradiated with ultra-violet light,[59] it was later discovered that vitamin D could be produced by irradiating the compound,

[57] Mellanby, E. (1918), *J. Physiol.*, 52, XI, iii.
[58] Chick, H. *et al.* (1923), *Spec. Rep. Ser. Med. Res. Coun. Lond.*, No. 77.
[59] Steenbock, H. (1924), *Science*, 60, 224.

ergosterol,[60] derived from yeast or, alternatively, the compound 7-dihydrocholesterol[61] prepared from cholesterol itself obtained from pigs' skin. This led to the elucidation of at least two forms for yet one further vitamin. The chemical structure of these two versions of vitamin D are shown in Figure 5.

The discovery that vitamin D activity could be produced by irradiating ergosterol was of major importance. Although ergosterol is not a particularly common substance, neither is it rare, and supplies were not difficult to procure; yeast was the most readily available source. Vitamin D_2, therefore, quickly became cheap and readily available. Initially, the production of vitamin D_2 by irradiation was a patented process, but the so-called 'Steenbock patent' was assigned to the alumni of Wisconsin University, who derived substantial sums of money from it which could be devoted to scientific and charitable purposes. The vitamin D_2 was admixed with a number of foods, of which baby foods were probably the most important. Up till this time some mixtures for artificial infant feeds, while causing the babies to grow and even win prizes at 'baby shows', had tended to give them rickets.

The availability of synthetic vitamin D made possible the preparation of infant foods amply supplied with the needs of the growing child. Capsules could be prepared for expectant mothers who had often found those other vehicles of the vitamin, fish-liver oils, distasteful. Vitamin D could also be provided at low cost as an ingredient of poultry feed to provide the special needs of chickens for their own bone growth and for the laying down of eggshell.

The very fact of its ready availability allowed public-health authorities to set standards for the composition of 'national dried milk' for infants and manufacturers to enrich their proprietary preparations. This, and the practice adopted by the producers of a variety of foods of irradiating their products with ultra-violet light, bringing about some measure of vitamin D synthesis *in situ*, led to a less favourable result than the one intended.

In 1952, Lightwood,[62] working at St Mary's Hospital in London, described a new disease which he called idiopathic hypercalcaemia. It occurred in infants between the ages of 5 and 8 months. The

[60] Pohl, R. (1927), *Nactr. Geo. Wiss. Gottingen Math. phys. Klasse III*, 185.
[61] Wendaus, A. (1935), Lettre, H., and Schevk, F., *Ann.*, 52, 98.
[62] Lightwood, R. (1952), *Proc. Roy. Soc. Med.*, 45, 401.

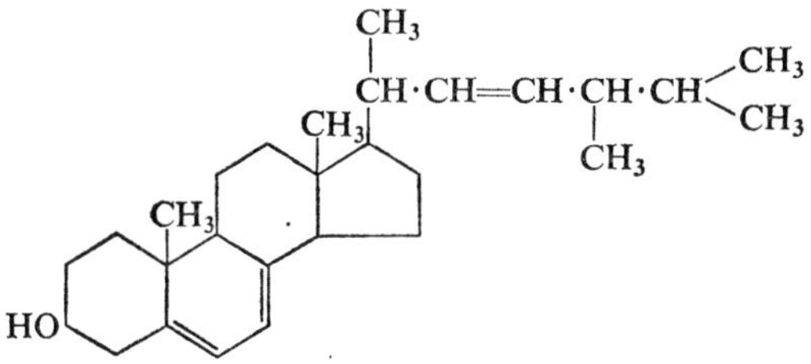

Ergosterol

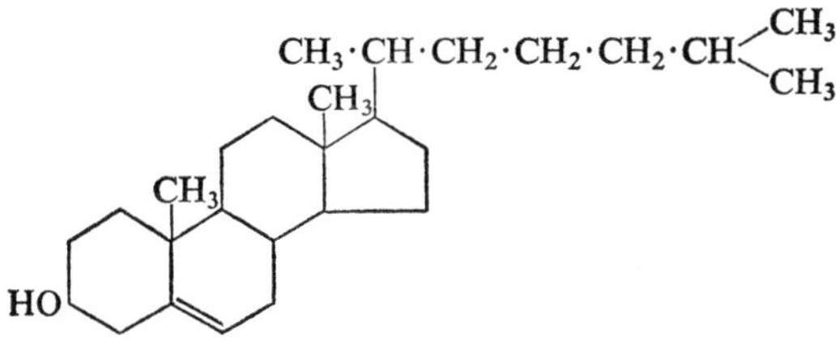

Cholesterol

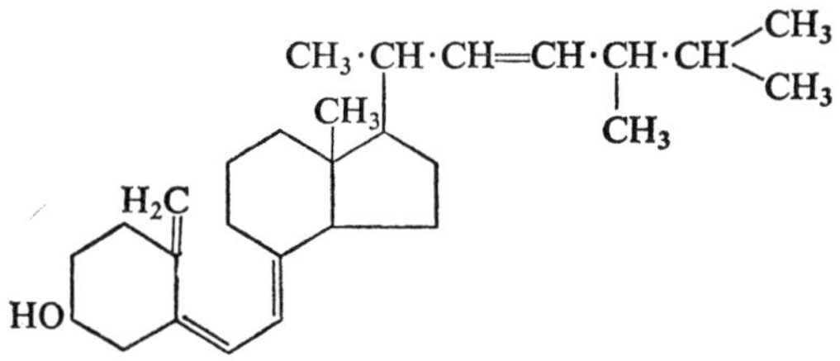

Calciferol (Vitamin D_2)

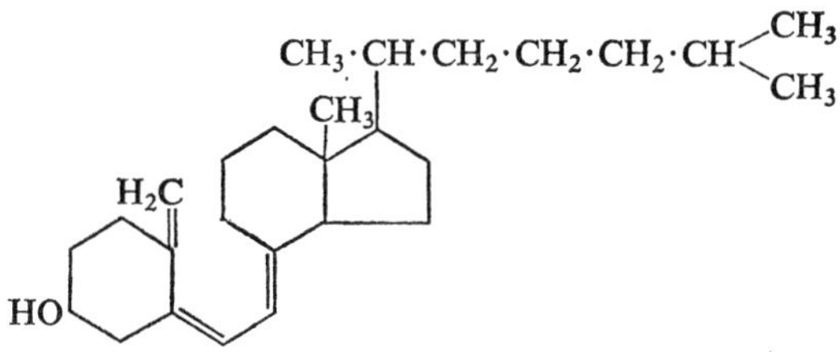

Activated 7-dehydrocholesterol (Vitamin D_3)

Figure 5 The chemical configuration of calciferol (vitamin D_2) and of ergosterol, the substance from which it is derived by irradiation; and of activated 7-dehydrocholesterol (vitamin D_3) and cholesterol

children refuse to eat, they vomit and develop a characteristic wasted appearance. This condition was subsequently recognised in several European countries.[63] Although the infants contracting it may have

[63] Forfar, J. O. (1959), and Tompsett, S. L., *Adv. Clin. Chem.*, 2, 167.

possessed certain predisposing characteristics, it was generally concluded that the illness was related to the consumption of excessive amounts of vitamin D. In 1957 the authorities responsible in Great Britain for the distribution of welfare supplies of enriched dried milk for infant feeding recommended[64] that its vitamin D content be reduced, together with that in infant cereal preparations.

And so it can be seen that the chemical synthesis of vitamins is not only a practical possibility, but a matter of practice. In a situation where natural foods are available, a diet containing adequate supplies of the different vitamins needed can be obtained by normal adults, although under special circumstances, for infants receiving artificial

Table 10

*The amounts and prices of different vitamins synthesised in the USA**

	1955		*1960*		*1965*	
	1000 lb	*\$/lb*	*1000 lb*	*\$/lb*	*1000 lb*	*\$/lb*
Vitamin A (alcohol and esters)	170	98·34	516	62·55	677	28·14
Vitamin B_1 (thiamine)	234	39·16	328			
Riboflavin	311	30·08	525	12·09	958	10·37
Niacin	2318	3·24	3015	2·16	2717	1·64
Pyridoxine (vitamin B_6)	30	193·39	73			
Vitamin B_{12} (cyanocobalamin)	0·5	41 000·00	1	10 059·00	1	8 324·00
Pantothenic acid	555	12·22	927	4·41	1610	2·26
Vitamin D_2	2	426·00	1		1	180·00
Vitamin D_3	1	362·00	2	237·00	2	260·00
Vitamin E					507*	19·48
Vitamin K					156*	13·62
Vitamin C (ascorbic acid)	2354	6·97	5275	3·30	8629	1·97

* 1966 figures.

(Data derived from U.S. Tariff Commission, *Synthetic Organic Chemicals. U.S. Production and Sales.*)

[64] Central and Scottish Health Services Councils (1957), *Rep. Jnt. Sub-Cttee. on Welfare Foods.* HMSO.

feeding or under particular clinical conditions, synthetic vitamins can be of value. But there are many circumstances in real life where people do not select a proper diet. Some people like to indulge in soft drinks and do not like fruit or vegetables. For these, a soft drink containing synthetic vitamin C is better than one without it. Others may, for one reason or another, eat excessive amounts of white bread. For these, bread containing synthetic thiamine and niacin can be nourishing.

In the future, there may be circumstances when food manufactured by chemical synthesis may be marketed. When that time comes, the vitamins which must constitute a part of the synthetic foods can be produced and, indeed, are being produced in increasing amounts, and sold at what is usually a steadily diminishing economic price, as Table 10 shows.

CHAPTER 6

SYNTHETIC ALCOHOL

Sugars, the basis of most of all the foods eaten by man, can be synthesised from formaldehyde, and formaldehyde, readily made from petroleum or natural gas, is comparatively inexpensive. Large tonnages are used for manufacturing phenol–formaldehyde plastics—Bakelite and the like. Yet sugars are, in fact, *not* made from formaldehyde. On the contrary, great quantities of cereal, sugar and molasses, a by-product of the refining process containing 50% of sugar, have been used as a basis for manufacturing alcohol. This alcohol has been used as a chemical intermediary in industry or as the principal pharmacological ingredient in alcoholic beverages.

In the immediate past, a process for manufacturing synthetic alcohol has been worked out, and, in contrast to sugar, synthetic alcohol is now produced on a large industrial scale, and significant amounts of food have thereby become available for human or animal consumption. The alcohol prepared by synthesis—and here I am referring to ethanol—

$$\begin{array}{ccccc} & H & & H & \\ & | & & | & \\ H- & C & - & C & -OH \\ & | & & | & \\ & H & & H & \end{array}$$

is identical with that previously derived from fermentation, and fermentation alcohol and synthetic alcohol are used indiscriminately in chemical and industrial processes. Yet, no steps have been taken in any country to substitute synthetic for fermentation alcohol as an ingredient in beverages, although an equal or greater amount of food could be released for eating if this were done, as has been released by the use of synthetic industrial alcohol.

The separation of "spirit" from fermented liquor by distillation has been carried out since times of antiquity, but the first preparation of substantially pure ethanol was only achieved in 1796.[1] Nicolas de Saussure is credited with having determined its composition, although

[1] Lowitz, A. (1796), *Crell's Chem. Ann.*, 1, 1.

it was Sir Edward Frankland who established its structure. The first chemical synthesis of ethanol was carried out in 1825 by the reaction of ethylene gas with sulphuric acid.[2] A patent based on this reaction was issued in 1961.[3] The serious manufacture of alcohol from ethylene derived from petroleum began on a commercial scale in 1930, and by 1945 the output of synthetic alcohol reached about 580 000 000 gallons in the United States alone.[4]

The process by which most of the alcohol manufactured synthetically is produced is still based in principle on the reaction discovered in 1825. This so-called 'indirect hydration' or 'esterification–hydrolysis process' involves the following chemical reactions. Ethylene gas, separated during the process of petroleum refining, is pumped into a reaction chamber containing strong sulphuric acid at a concentration of 95–98%. A pressure of 150–200 lb per sq. in. is maintained and the temperature held at 50–80°C. Since the reaction produces heat, it is necessary to cool the mixture while the process is in operation. The changes that occur are that the ethylene

```
H H
| |
C=C
| |
H H
```

becomes converted into a mixture of monoethyl sulphate

```
    H H
    | |
H—C—C—O—SO₃H
    | |
    H H
```

and diethyl sulphate:

```
    H H
    | |
H—C—C
    | | \
    H H  O  O
          \ ‖
           S
          / ‖
    H H  O  O
    | | /
H—C—C
    | |
    H H
```

[2] Faraday, M. (1825), *Phil. Trans. Roy. Soc.* (*London*), 115, 440.
[3] U.S. Patent 41,685; Cottelle, E. A. (1862), *Bull. soc. Chim.*, A4, 279.
[4] Caldwell, D. L., and Lichtenstein, I., in Kirk-Othmer (ed.) (1965), *Encyclopaedia of Chemical Technology*, vol. 8, 422.

These mixed sulphates are hydrolysed by being allowed to stand with something between an equal amount of water to a double amount. The precise conditions of concentration, time and temperature are important since, besides freeing the ethanol—that is, 'alcohol'—from the sulphate, a significant quantity of ether is also formed. The formation of this highly flammable and toxic substance can be minimised by exact control of the process. The chemical reactions are as follows:

$$H{-}CH_2{-}CH_2{-}O{-}SO_3H + HO{\cdot}H \longrightarrow H{-}CH_2{-}CH_2{-}OH + H_2SO_4$$

$$\begin{array}{l} H{-}CH_2{-}CH_2{-}O \\ H{-}CH_2{-}CH_2{-}O \end{array}\!\!\Big> SO_2 + \begin{array}{l} H_2O \\ H_2O \end{array} \longrightarrow \begin{array}{l} H{-}CH_2{-}CH_2{-}OH \\ H{-}CH_2{-}CH_2{-}OH \end{array} + H_2SO_4$$

$$\begin{array}{l} H{-}CH_2{-}CH_2{-}O \\ H{-}CH_2{-}CH_2{-}O \end{array}\!\!\Big> SO_2 + H{-}CH_2{-}CH_2{-}OH \longrightarrow H{-}CH_2{-}CH_2{-}HSO_4 + \underset{\textit{(diethyl ether)}}{H{-}CH_2{-}CH_2{-}O{-}CH_2{-}CH_2{-}H}$$

The alcohol produced by these reactions is recovered by distillation. Ether is separated from it by steam-distillation. The purification of the alcohol may, in fact, be quite complicated. For a start, the highest concentration obtainable is 95–96%, the remainder being water. Of minor contaminants, acetaldehyde may be the most troublesome to remove. This compound is derived from any acetylene which may be present in trace quantities in the original ethylene feed-stock. If, as sometimes happens, acetone or acetaldehyde are present in the feed, the quality of the alcohol produced will again be affected. Propylene and butadiene may also occur mixed with the starting ethylene. They lead to the production of polymers. It is clear from all this that the control of the synthetic process, simple as it may seem, presents many difficulties. Soviet workers[5] have used conditions differing

[5] Noyes Development Corp. (1962), *Ethyl Alcohol Production Technique*. Pearl River, N.Y.

somewhat from those employed in the United States. Other workers have used catalysts[6] to accelerate the reaction as the ethylene is bubbled in an upward stream through the downward-flowing sulphuric acid in the pressure vessel. The process is a practical one, and large amounts of alcohol have been made by it. Nevertheless, to obtain pure alcohol has been, and remains, a difficult problem.

In 1951, an even more direct process for producing alcohol from petrol was patented.[7] This process was introduced into commercial operation in 1954. The main reaction, which is brought about by the use of an appropriate catalyst, could hardly be simpler. It is this:

$$\underset{\text{Ethylene}}{\mathrm{H_2C{=}CH_2}} + \underset{\text{Water}}{\mathrm{H_2O}} \rightarrow \underset{\text{Ethanol}}{\mathrm{H{-}CH_2{-}CH_2{-}OH}} + \text{a large amount of heat}$$

The manufacture of alcohol by the direct hydration of ethylene has been successfully operated by several big companies. Shell discovered in 1948 that phosphoric acid could be used by itself as a catalyst, and it was this firm that patented its use in 1951. Eastman Kodak[8] and Union Oil[9] patented the use of modified catalysts in which the phosphoric acid was incorporated with metallic carriers. The Italian firm of Montecatini devised in 1965 a continuous process in which the ethylene at a temperature of 290°C and a pressure of 5100 lb per sq. in. is brought into contact with a catalyst composed of aluminium hydroxide.[10]

There are many other chemical methods known by which alcohol can be synthesised. For example, other hydrocarbons than ethylene can be used as a starting material. Among these are propane and butane, acetaldehyde or even diethyl ether,[11] often considered to be the most inconvenient contaminant of the more practicable processes.

[6] Ellis, C. (1937), *Chemistry of Petroleum Derivatives*, II. Reinhold, N.Y.
[7] U.S. Patent 2,579,601.
[8] U.S. Patent 2,773,910.
[9] U.S. Patent 2,756,247.
[10] U.S. Patent 3,164,641.
[11] Kozlov, N. S. (1936), and Goluborskaya, N., *J. Gen. Chem., U.S.S.R.*, 6, 1506.

Alcohol can even be synthesised directly from carbon monoxide gas and hydrogen.[12]

The manufacture of alcohol by chemical synthesis using petroleum fractions as starting material clearly represents a significant example of the production of food by a direct manufacturing process. To take the first aspect of the operation first, the substantial tonnages of cereal grains and molasses used for the manufacture of industrial alcohol before the synthetic methods were available were released for human and animal consumption. There is, however, a second aspect of the matter which needs to be discussed at some length in any consideration of food synthesis. This is the matter of taste and smell. It is true that foods are composed of the several nutrients which they contribute to the metabolism of the body. But, as we have already seen from the chapter on synthetic vitamins, even when it can justly be claimed that the chemical structure of one or other of these is known—as was the case with vitamin A, for example—later study may show that in nature vitamin A may exist in half a dozen or more similar, but different, chemical configurations. But beyond the separate and fairly well understood nutrients, there are in natural food a very large number of chemical compounds present often in only trace amounts, but which nevertheless contribute to their 'quality'. These may be part of what we loosely describe as 'freshness', they constitute bouquet and delicacy of taste and, for all we know to the contrary, to nutritional value as well.

Synthetic alcohol is every bit as useful an organic chemical as is alcohol produced by fermentation. There is, however, reason for the general refusal of those who manufacture alcoholic beverages to decline to use it as an ingredient of mixtures intended to be drunk. The next two chapters deal with aesthetic topics, with taste, smell and colour, and with consistency and general appearance. It is, however, useful at this stage to review the complexity of the processes of biological chemistry by which alcohol is formed in nature, to show how diverse are the trace compounds which, in larger or lesser concentration, accompany beverage alcohol and hence contribute to its 'character', and compare it with synthetic alcohol which, if it contains any trace of impurity at all, will contain residues of the petroleum feed-stock from which it came.

Detailed studies of the biochemical mechanism by which sugar

[12] Aries, R. S. (1947), *Chem. Eng. News.*, 25, 1792.

is converted into alcohol during fermentation by yeast showed that it proceeds in a number of stages. The first of these brings about a coupling of the glucose molecule with a phosphate radical,

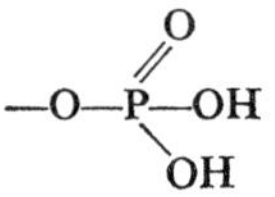

```
       O
      //
—O—P—OH
      \
       OH
```

The next causes the intra-molecular rearrangement by which the glucose phosphate ester is converted into a fructose phosphate ester. A further enzyme couples on a second phosphate group, producing the Harden–Young ester, so-called in honour of the two chemists who identified it. And so the series of changes continues, as shown below, until the substance, 1,3-diphosphoglyceraldehyde is produced.

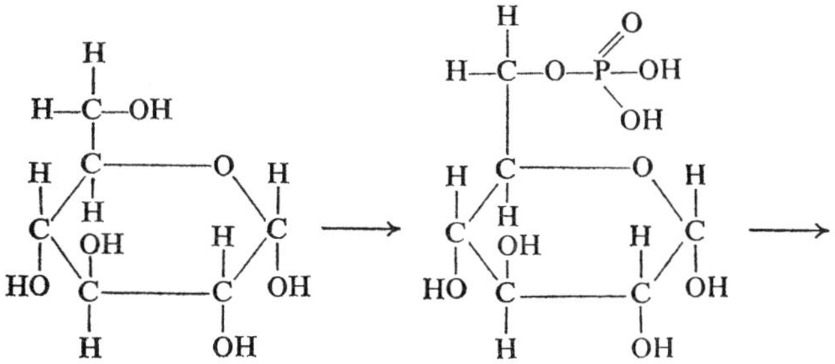

Glucose → *Glucose phosphate* →

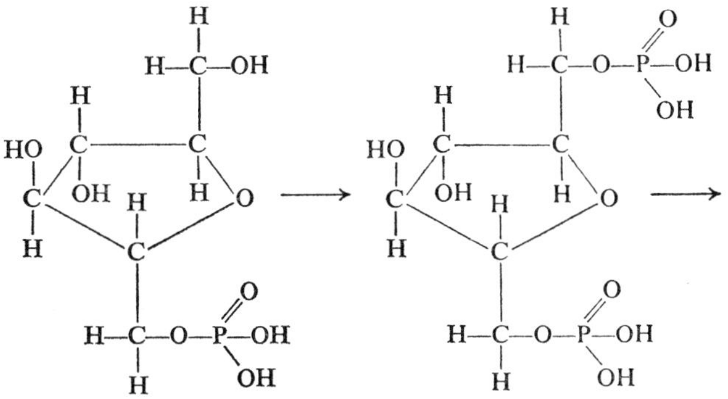

Fructose phosphate → *Harden-Young ester* →

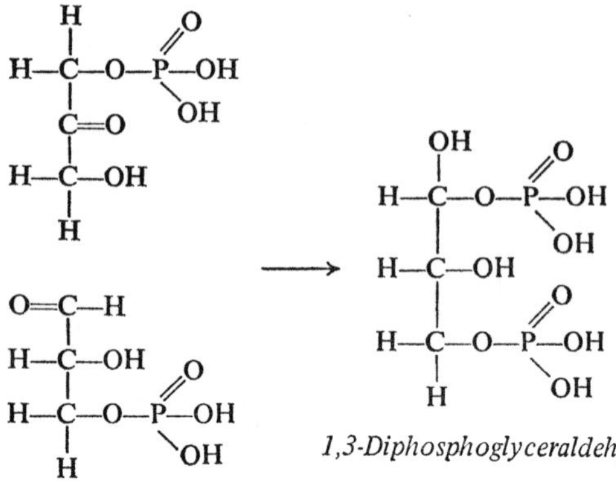

1,3-Diphosphoglyceraldehyde

The compound 1,3-diphosphoglyceraldehyde plays a central part in the chain of linked reactions by means of which part of the chemical energy in the original glucose molecule is released in the controlled and continuous manner required for the biological activities of the creatures in whose cells the mechanism of fermentation is taking place. The 1,3-diphosphoglyceraldehyde, under the influence of an enzyme complex, acts as a kind of chemical 'pump'. It becomes *oxidised* and by this fact causes a molecule of acetaldehyde which is formed later on to become *reduced.* And its reduction is alcohol. The compounds concerned are the following:

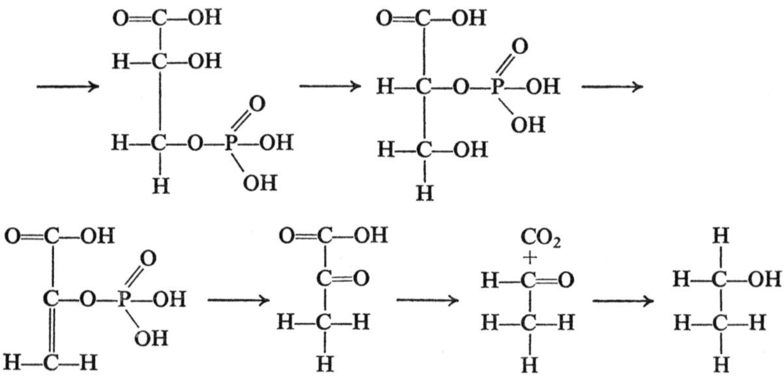

Pyruvic acid → CO_2 + *Acetaldehyde* → *Ethanol*

This whole process is sometimes called the Embden–Meyerhof–Parnas system, in honour of three of its principal architects. The system operates in a state of dynamic balance. If the acidity of the fermentation medium falls, a proportion of glycerol is produced from the 1,3-diphosphoglyceraldehyde and the amount of alcohol diminishes. If the concentration of dissolved oxygen increases, another cycle of changes begins to operate. This is the so-called 'tricarboxylic acid cycle'—also called the 'Krebs cycle', again after its principal author, Hans Krebs—which takes over at the stage of pyruvic acid (shown in the diagram on p. 84). This involves the cyclic formation of citric acid, α-ketoglutaric acid, succinic acid, fumaric acid and oxaloacetic acid. This series of changes is concerned with the growth of the fermenting yeast. And at the same time that it grows, other sequences of change lead to the formation of the so-called higher alcohols, propanol

$$\begin{array}{ccccccccc} & & \mathrm{H} & & \mathrm{H} & & \mathrm{H} & & \\ & & | & & | & & | & & \\ \mathrm{H} & - & \mathrm{C} & - & \mathrm{C} & - & \mathrm{C} & - & \mathrm{OH} \\ & & | & & | & & | & & \\ & & \mathrm{H} & & \mathrm{H} & & \mathrm{H} & & \end{array}$$

n-butanol

$$\begin{array}{ccccccccccc} & & \mathrm{H} & & \mathrm{H} & & \mathrm{H} & & \mathrm{H} & & \\ & & | & & | & & | & & | & & \\ \mathrm{H} & - & \mathrm{C} & - & \mathrm{C} & - & \mathrm{C} & - & \mathrm{C} & - & \mathrm{OH} \\ & & | & & | & & | & & | & & \\ & & \mathrm{H} & & \mathrm{H} & & \mathrm{H} & & \mathrm{H} & & \end{array}$$

and *n*-pentanol

$$\begin{array}{ccccccccccccc} & & \mathrm{H} & & \mathrm{H} & & \mathrm{H} & & \mathrm{H} & & \mathrm{H} & & \\ & & | & & | & & | & & | & & | & & \\ \mathrm{H} & - & \mathrm{C} & - & \mathrm{C} & - & \mathrm{C} & - & \mathrm{C} & - & \mathrm{C} & - & \mathrm{OH} \\ & & | & & | & & | & & | & & | & & \\ & & \mathrm{H} & & \mathrm{H} & & \mathrm{H} & & \mathrm{H} & & \mathrm{H} & & \end{array}$$

These may occur in various alternate configuration with branched in place of straight carbon chains. The group of higher alcohols taken together is also designated as 'fusel oil'.

Yet another pathway of change which may take place to a greater of lesser degree during the fermentation of sugars to produce alcohol is the 'pentose phosphate cycle', also designated as the 'Calvin cycle'. This series of changes is practically identical with the photosynthetic cycle, except that it serves to release energy to the living yeast, whereas in leaves it converts the light energy of the sun into the chemical energy which the yeast acquires.

The relevance of this information in a discussion of the synthesis of alcohol is that it makes clear the fact that in the production of alcohol by biological fermentation numerous substances take part, and small concentrations of them may be present as trace contaminants in the alcohol when it is subsequently separated.

In 1960, the approximate cost of a gallon of 95% alcohol produced in the United States from molasses by fermentation was 45 cents.[13] This price was, however, subject to wide fluctuation due to the large variation in the price of sugar which is traditionally unstable. The cost of a gallon of alcohol made from maize, when malt was used as the source of sugar was at this same date approximately 70 cents. Here again, fluctuation in the price of grain could drastically affect the price of alcohol. Other calculations based on raw-material cost, the cost of labour, plant and the price for which by-products, including surplus yeast derived from the fermentation, could be sold, gave figures of 30 cents a gallon for alcohol made from waste sulphite liquor and from acid-hydrolysed wood. At about the same date, the American firm Union Oil and the Italian Montecatini estimated that alcohol could be produced from a petroleum fraction for a figure less than, or at least not in excess of, 30 cents per gallon.

Between 1950 and 1960, the total production of industrial alcohol in Great Britain remained substantially steady at about 35 000 000 gallons a year.[14] But during this decade, the proportion of the total gallonage made up of synthetic alcohol rose from about 8% in 1950 to 70% in 1960. The changes in production in the United States can be followed in more detail,[15] and demonstrate the extent to which synthetic alcohol has replaced fermentation alcohol. In 1930, of the 102 000 000 gallons manufactured, 96 000 000 were made from molasses. Ten years later in 1940, the total was 128 000 000 gallons, of which 88 000 000 were from molasses, 8 000 000 from grain and the remaining 32 000 000 synthesised by the esterification–hydrolysis

[13] Chilton, C. H. (1960), *Cost Engineering in the Process Industries*. McGraw-Hill, N.Y.

[14] Carle, T. C. (1962), and Stewart, D. M., *Chem. & Ind.*, 830.

[15] U.S. Bureau of Internal Revenue.

process. By 1950, the total was 165 000 000 gallons, only 57 000 000 were derived from molasses, 4 000 000 from grain, sulphite liquor and other sources combined. Of the rest, 90 000 000 gallons were synthesised by the esterification process and 14 000 000 by direct hydration. When the figures are followed up to 1960, of the total of 273 000 000 gallons, 25 000 000 were produced by fermentation from all the different raw materials used and 248 000 000 gallons were synthesised, 42 000 000 by the method of direct hydration.

On average, a gallon of alcohol can be produced by fermentation from 20 lb of molasses, equivalent to about 10 lb of sugar, or from 15 lb of maize. It can readily be appreciated that very large amounts of potentially edible human and animal food were, before the advent of the synthetic process, devoted to the manufacture of a 'chemical' —ethanol—much of which was used as a basic raw material in chemical industry, or for such diverse purposes as automobile anti-freeze, adhesives, tooth-paste, shellac, explosives or (according to American statistics) embalming fluid. For example, in 1938, the estimated world production of industrial alcohol, almost all of which was produced by fermentation, was 940 000 000 gallons. This can be calculated to be equivalent to approximately 6 500 000 tons of maize. Here, then, we see an example of a synthetic process making, in practice, a major contribution to the world's food supplies.

This contribution is, however, indirect. It arises, also, not from a considered attempt to make a food-stuff by artificial means, but from the use of technological innovation to improve the economic efficiency of an industrial process. I have argued already in this book that it is mistaken to assume that food possesses claims quite different from those of other economic commodities merely because communities must eat to live. This is indeed true, and there may be circumstances under which food has absolute priority and must be provided as a matter of 'welfare' rather than as merely one among many of the items of economic wealth demanded by a nation's population.

Alcohol is a food. A gram of alcohol contributes 5·7 calories of energy to the body compared with the 4 calories derived from 1 g of sugar. Of course, it is only in very special circumstances that alcohol is consumed for its food value. In general, it is used partly as a delicacy and partly as a social analgesic. Although the pharmacological activity of alcohol is the attribute which, above all others, gives it its attractiveness as an article of internal consumption, its character, that is to say its taste and smell, is a matter of importance as well. Although in certain extreme circumstances people will drink alcoholic beverages of almost any composition so long as they contain alcohol, and will, therefore, provided they are consumed in sufficient quantities, make those who take them drunk, under more normal circumstances great importance is attached to the detailed composition of alcohol-containing beverages.

The purification of ethanol is a difficult technical undertaking. Ethanol is not only completely miscible with water, but is also equally miscible with many organic compounds as well. Elaborate procedures are required to obtain ethanol of high purity. Repeated distillation from aqueous solutions of various strengths or from solutions of non-volatile solutes of various concentrations is used. Further procedures may include the treatment of the alcohol with charcoal or activated carbon to adsorb, and subsequently remove, traces of organic contaminants. But, even so, great difficulty may be encountered in freeing the alcohol from every trace of contaminant, or, indeed, in reducing the residual traces of contaminants to a level sufficiently low so that no evidence of their presence can be detected by smell. In the preparation of beverage alcohol, repeated distillation so managed as to yield a final concentration of 95% ethanol—the remainder being water—gives what is designated 'neutral spirit'. But in spite of this quite elaborate purification, it is often possible to recognise by the smell of the alcohol the raw material from which it was derived. Even more elaborate precautions are often taken to obtain alcohol of an even higher degree of purity for use as a solvent in the blending of perfumes.

Neutral spirit is used in the manufacture of such beverages as gin and vodka. The flavour of gin is provided by the extraction of a mixture of vegetable products of diverse sorts—so-called 'botanicals' —which is added to gin spirit. The particular character of vodka is due to the absence of almost every smell except that of pure ethanol. It is interesting to find that although neutral spirit derived from the fermentation of grain, of molasses, and, at least for the manufacture of vodka, of potatoes as well, has been employed in the manufacture of these beverages, the alcohol used is always produced by fermentation.

There are three reasons why synthetic alcohol has not been used as a component of drink, even though, were it so employed, substantial amounts of food would become available. The first reason is because of the tax laws which exist in most of the advanced countries where synthetic alcohol is made or where it is available as an article of commerce. Duty is levied on alcoholic beverages, but, at the same time, there is no wish to hamper the chemical industry by restricting the use of a versatile chemical intermediary. It therefore happens that synthetic alcohol is commonly kept separate from those places where beverages are manufactured. A second, and more substantial, reason for denying the use of synthetic alcohol to the beverage trade is a fear of possible toxicity. This possibility can be considered remote in view of the high degree of chemical purity of the synthetic material. Perhaps of all the reasons adduced, the third carries the most practical weight. It is the difficulty with synthetic alcohol, as with fermentation alcohol, of freeing it from the last traces of taste and smell.

People do not ingest nutrients and obtain their calorie requirements, they eat the foods they enjoy and to which they are accustomed and satisfy their appetites.[16] It is equally true to say that they do not drink alcohol, they select wines and discriminate quite subtle differences between one kind and another. And their choice of different kinds of spirits is equally sophisticated. To describe whisky as a beverage containing from 35 to 40% of alcohol is to miss the qualities that equally influence the consumers who drink it. Connoisseurs of wines accept or refuse different vintages on the basis of subtle differences in composition reflected in even more delicate nuances of taste—and support these sensory emotions on the hard touchstone of economics. So too, a major international export industry (the trade in Scotch whisky), covering very substantial sums of money and large tonnages of edible cereals is entirely dependent on the presence mixed with the ethanol (the main component of the beverage) of many organic compounds present only in trace amounts and derived from the complex mechanism of the fermentation process. The presence of several hundred of these trace components has been detected,[17] although only 20 or 30 have been identified. And the contribution of few of these to the taste and smell of whisky is properly understood.

The complexity of what might be thought to be a comparatively simple beverage, such as unmatured malt whisky, the main com-

[16] Pyke, M. (1968), *Food and Society*. Murray, London.
[17] Potterat, M. (1966), and Gygyi, R., *J. Anal. Chem.*, 231, 312.

ponents of which are, say, 40% of alcohol (that is, ethanol) and 60% of water, is shown in Table 11.

Table 11

The volatile components other than ethanol detected by gas-chromatography in unmatured malt Scotch whisky[18]

Component	*Concentration present in p.p.m.*			
	1	2	3	4
Acetaldehyde*	39	37	37	38
Diethyl acetal	14	14	17	17
Acetone	<0·2	<0·2	<0·2	<0·2
Isobutyraldehyde	15	14	15	15
Isovaleraldehyde	20	21	22	21
n-Propanol	320	280	270	280
Isobutanol	530	510	490	500
n-Butanol	7	6	5	5
Isopentanol	1680	1620	1570	1560
Ethyl acetate	420	440	450	440
Isopentyl acetate	32	31	28	28
Ethyl lactate	10·6	11·0	9·9	9·7
Ethyl caprylate	18·8	18·0	18·1	17·8
Ethyl caprate	56·7	58·2	54·4	56·5
Ethyl laurate	49·5	45·0	42·8	43·1
Ethyl myristate	3·9	3·8	3·5	3·4
Ethyl palmitoleate	29·6	28·5	26·2	25·4
Ethyl palmitate	39·7	38·1	35·4	33·7
Furfural	12·4	12·2	12·6	12·4
β-Phenyethyl acetate	20·9	19·1	18·7	18·0
β-Phenylethanol	31·8	29·6	27·1	28·2

* Includes acetaldehyde combined as acetal.

Large areas of good agricultural land are used to grow barley for whisky, just as are extensive areas employed to grow grapes for making wine and brandy. The analytical figures in Table 11 are paralleled for brandy, rum and the different kinds of wine, all of which depend for their final flavour and smell on components produced simultaneously (at least in the unmatured stage of manufacture) with the production of the alcohol. For the chemical manufacturer, as for the nutritionist anxious to contribute to the world's

[18] Ducan, R. E. B. (1966), and Philip, J. M., *J. Sci. Food Agric.*, 17, 208.

food supply, the problem of producing an equally attractive beverage in which synthetic alcohol prepared from, say, natural gas, is difficult, if not insuperable. On the other hand, several widely consumed beverages, of which gin in its numerous varieties and vodka are two, are prepared from highly purified ethanol to which—at least in the manufacture of gins—various vegetable extracts, themselves made by treating the vegetable products with the purified alcohol, are subsequently mixed. There would seem to be no technical objection to making these beverages from synthetic alcohol.

The subtlety of the composition of accepted alcoholic beverages with which drinks compounded with synthetic alcohol would have to compete becomes even more apparent when it is remembered that such distilled drinks as whisky and brandy, quite apart from non-distilled wines, are commonly matured for long periods of time, usually extending over years, before they are drunk. The expense of providing containers and warehouses in which to store them, the cost of transport and handling, as well as the inevitable loss by evaporation and leakage, are a measure of the importance to be measured in the solid, economic terms of large sums of money, attached to what can only be called a trivial change in chemical composition. This change, however, is accompanied by a change in taste and smell which can only be accepted as favourable, since it is the purpose for which the whole operation is undertaken.

To bring about this change in whisky, for example, the liquor is stored in oak casks; no other type of vessel and no other kind of wood are accepted. Evidence has been published[20] to show that the actual substance of the wood enters into chemical combination with the alcohol content of the whisky and undergoes a process of so-called ethanolysis. It is truer than might be imagined to say that a man has drunk a cask of whisky. It follows, therefore, that mature whisky contains non-volatile components including tannins and polyphenolic extractives as well as the products derived from the chemical disintegration of the oak lignin. The extent to which these contribute to the flavour of whisky is not clear. Nor are the changes that take place in the volatile components during the ageing process at all well understood, in spite of the considerable research which has been carried out. The economic charge represented by the tedious and ostensibly inefficient process of maturation has been recognised by the major scientific efforts deployed in an attempt to shorten the time found necessary to produce what, in commercial terms, can be accepted as a satisfactory product. Singleton[19] in 1962 cited some 400 references to work carried out in different parts of the world aimed at reducing the duration of the maturing process. It is generally agreed that none of these investigations has been successful. While this is so, it is doubly doubtful whether synthetic alcohol will be accepted as an ingredient of such drinks.

As the techniques of chemistry and physics advance, so are the biological complexities of fermented alcoholic beverages revealed to be more and more extreme. It almost seems as if advances in chemical analysis have outstripped understanding of the physiology of taste and smell. It is this perception of taste and smell, which to the nutritionist concerned with malnutrition and poverty seems almost frivolous, and the social desire for alcoholic beverages at all which are the areas of biology about which scientists know least. And among such scientists are those concerned with the synthesis of food. Consider Table 12. Here we have a list of compounds identified in rum. In carrying out these identifications, Maarse and de Brauw[20] employed sophisticated techniques of chemical analysis. For what purpose? it may be asked. The answer, however, is fundamental to any serious study of food science, since food, whether of biological origin, as it has all been up to now, or of synthetic manufacture, as scientific possibility now make feasible, is designed to be eaten—or drunk.

The main thesis of this book is to demonstrate that food can be made by the artificial means of chemical synthesis. The essential feature of food comprises the nutrients of which it is composed. The life of a man is maintained because he consumes an adequate amount of protein, carbohydrate, fat, mineral components and vitamins. All of these, as has been described, can be produced by chemical synthesis. The fact that this is so may, however, be of little practical significance—for two reasons. First, because the nutrients synthesised, as is the case with sugar, are very much more expensive than the same material produced by the normal biological means, that is to say, naturally. The second reason may be that the synthetic material is unattractive to the consumer. Both of these are valid reasons. Price is determined by a variety of causes, principally the strength of the

[19] Singleton, V. Z. (1962), *Hilgardia*, 32, 319.
[20] Maarse, H. (1966), and de Brauw, M. C. T., *J. Food Sci.*, 31, 951.

Table 12

The components identified in rum

Ethyl formate	2-Methylfuranid-3-one
Ethyl acetate	Ethyl pentanoate
Diethoxymethane	3-Methylbutan-1-ol
2-Methylbutanol	Ethyl hexanoate
1,1-Diethoxyethane	Furfural
Ethyl propionate	2-Acetylfuran
n-Propyl acetate	Ethyl heptanoate
n-Propanol	Benzaldehyde
Ethyl isobutyrate	5-Methylfurfural
Isobutanol	Ethyl octanoate
Ethyl butanoate	Ethyl benzoate
Ethyl 3-methylbutanoate	Diethyl succinate
n-Butanol	Ethyl nonanoate
3-Methylbutyl acetate	

desire for the article felt by the purchaser. Nutritionists and philanthropists deceive themselves if they assume either that food is the commodity which people desire above all else or that people select the food they eat on the basis of the nourishment they get from it.

Alcoholic beverages represent a striking example of the operation of the principles that I have just described. The barley grown by a farmer can be eaten directly by people or it can be fed to livestock. Yet the kind of barley that commands the highest price—that is, the barley which consumers desire the most—is malting barley intended for brewing beer or distilling whisky. The pharmacological effect of these beverages, it can be argued, is due to the presence in them of an appropriate proportion of ethanol. Indeed, ethanol prepared from ethylene, derived from petroleum or natural gas by one of the synthetic processes I have described, will make a man drunk in exactly the same way as would an equal amount of alcohol (derived from grain, grapes or molasses) in gin. But, just as a customer at a butcher's shop will willingly pay handsomely for a prime joint of meat rather than buy for much less an equal quantity of nutrients in the form of mince, so will a descriminating drinker express his desire for a particular—and non-intoxicating—taste and smell by the very practical process of paying for it as a particular beverage.

During the Second World War, there was a shortage of petrol in Great Britain. In order to implement an equitable rationing system and ensure an equal distribution for all, the government mixed together all the shipments of fuel as they arrived and marketed a so-called 'pool petrol', issued, without favour, in exchange for coupons to every holder of a ration book. The success of this scheme excited the interest of the Ministry of Food, where a similar scheme was prepared to insure an equally just distribution of the scanty supplies of wine in the country. The scheme for a 'pool wine', however, never came about. The chemical differences between one wine and another may not be very great, but nevertheless the differences, minor though they may be in terms of nutrition and pharmacology, are considered to be important by the people who drink the wine.

The chemical components of wine which give it its special character are not fully understood. Sherry is, in the United States, a major article of commerce, where, perhaps unlike Spain, strenuous efforts are being made to elaborate a proper specification in chemical terms so that industrial production can be operated on a satisfactorily systematic basis. Yet only part of the problem has so far been dealt

Table 13

Acidic components identified in Spanish flor sherry

In large concentration
Caproic acid
2-Hydroxycaproic acid
2-Furoic acid
Succinic acid
Benzoic acid
2-Hydroxy-3-phenylpropionic acid
In moderate concentration
Lactic acid
3-Hydroxycaprylic acid
2-Hydroxyisovaleric acid
Caprylic acid
Phenylacetic acid
In small concentration
Glutaric acid
Azeloic acid
Tricarballylic acid
Vanillic acid

with. In a recent paper,[21] only the acidic components of two Spanish sherries were determined with any completeness, and even then the contribution made by the various compounds identified could not exactly be specified. The list is shown in Table 13.

The synthesis of alcohol undoubtedly represents a practical operation for making food from non-food sources. It can itself be consumed or its production serve to release land otherwise occupied in growing fermentable material. Yet the limitation in its use highlights the importance of the trace compounds whose function is limited merely to contributing smell and taste.

[21] Kepner, R. E. (1968), Webb, A. D., and Maggiosa, L., *Am. J. Enol. Vitic,* 19, 116.

CHAPTER 7

SYNTHETIC FLAVOURS, ODOURS AND COLOURS

Foods, to be acceptable, must possess an appropriate taste and smell, their appearance must be attractive and, as will be discussed in Chapter 8, they must also have a desirable consistency. It follows, therefore, that before synthetic foods can hope to appeal to consumers other than dedicated participants in scientific experiments or people in the last extremes of hunger, they must contain synthetic flavours as well as synthetic proteins, fats and carbohydrates. Synthetic flavourings have already been compounded. To be successful, such combinations must, in fact, appeal not only to what in physiological terms is strictly the sense of taste, but also to the sense of smell as well.

As was apparent in Chapter 6, the flavour of food is derived from a complex mixture of compounds. Physiologists, attempting a first approximation of the nature of taste as recorded by the 'taste buds', that is, the sense organs situated at the base of the tongue, have produced evidence to support the hypothesis that the taste of any one of all the variety of edible substances there are can be interpreted in terms of four components: sweet, sour, bitter and salty. If this were so, the production of synthetic taste to flavour synthetic food would be easy. It would only be necessary to mix together the right proportions of quinine, of which as little as 5 parts in 10 000 000 can be tasted as bitter, hydrochloric acid, HCl, of which 7 parts in

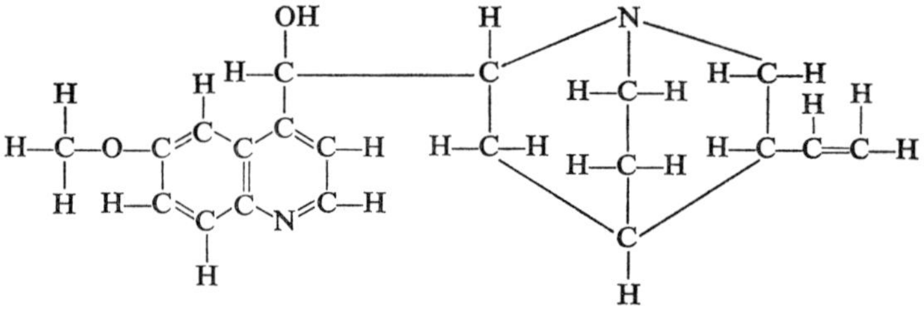

Quinine

100 000 can just be detected as sourness, with sodium chloride (salt), NaCl, of which 2 parts per 1000 can be tasted, and sugar, the chemistry of which was discussed in Chapter 5, of which 5 parts in 100 are detectable, for any taste under the sun to be reproduced. In fact, this is not so. To produce a tolerable imitation pineapple flavour, about ten pure chemical compounds, which can be synthesised without too much trouble, mixed with some seven natural oils and concentrates, which could not be synthesised without a great deal of trouble, if at all, are required. These are listed in Table 14.

Table 14

Artificial pineapple flavour[1]

1. *Pure compounds*	%
Allyl caproate	5·0
Isopentyl acetate	3·0
Isopentyl isovalerate	3·0
Ethyl acetate	15·0
Ethyl butyrate	22·0
Terpinyl propionate	2·5
Ethyl crotonate	5·0
Caproic acid	8·0
Butyric acid	12·0
Acetic acid	5·0
2. *Essential, oils, etc.*	%
Oil of sweet birch	1·0
Oil of spruce	2·0
Balsam Peru	4·0
Volatile mustard oil	1·5
Oil cognac	5·0
Concentrated orange oil	4·0
Distilled oil of lime	2·0

The complexity implicit in this table emphasises the basically intricate nature of taste. The chemical complication of natural flavours is reflected in the parallel multiplicity in the sense organs by which they are perceived. For example, whereas a chicken possesses 24 taste buds, a pigeon 37, a starling and a duck 200, a parrot 350, a kitten 473, a bat 800 and a dog 1706, a man has 9000. On the other

[1] Guenther, E. ed. (1966), *Kirk-Othmer Encyclopaedia of Chemical Technology*, vol. 9, 357. N.Y.

hand, a pig and a goat possess 15 000, a rabbit 17 000, a cow 25 000, while, leading all the rest, comes the catfish with 100 000 taste buds.[2] Yet the species with the greatest number of taste buds does not necessarily possess the highest degree of sensitivity and discrimination in tasting. A chicken with only two dozen taste buds will reject solutions on grounds of a taste which a cow, with 25 000 taste buds, apparently cannot detect at all.

Although delicate tastes such as those of natural food-stuffs—for example, pineapple, tomato, fresh bread or coffee—are made up of complex mixtures of chemical compounds contributing taste, together with many more providing smell, it is possible to categorise some of the principal kinds of tastes in terms of specific chemical molecules. So-called ionized salts of low molecular weight, of which 'common salt', NaCl, is the best known, have a salty taste. Besides sodium chloride, ammonium chloride, potassium chloride, lithium chloride, sodium bromide, ammonium bromide, sodium iodide and lithium iodide taste salty. As the molecular weight becomes greater, the salty taste becomes tinged with bitterness. For example, potassium bromide and ammonium iodide taste both salty and bitter, and caesium chloride, rubidium bromide and potassium iodide taste predominantly bitter.[3]

Sourness, like saltiness, is also, for the most part, reflected in terms of chemistry. Indeed, the relationship between acids and their acidity and the taste of sourness is so direct that part of the original definition of an acid was a chemical compound with a sour taste. The taste of sweetness, on the other hand, is a much more complex matter. Almost every class of organic compound possesses members which have a sweet taste. Lead acetate, for example, was at one time known as 'sugar of lead' on account of its taste, and instances of poisoning used to occur due to its being added to beer as a sweetener. Certain compounds have been discovered to possess enormously strong, sweet tastes, yet the relationship between chemical configuration and sweetness is by no means clear. The compound, 2-amino-4-nitropropoxybenzene (opposite) is 4000 times sweeter than sugar. Yet if the positions of the $—NO_2$ group and the $—NH_2$ group in the molecule are transposed, the new compound, 4-amino-2-nitropropoxybenzene, possesses no taste at all. Replacement of the

[2] Kare, M. R. (1966), *Agric. Sci. Rev.*, 4, 10.

[3] Kionka, L. (1922), and Stratz, A. O., *Arch. expl. Path. Pharm.*, 95, 241.

```
              H   H   H
              |   |   |
          O—C—C—C—H
          |   |   |   |
          |   H   H   H
          |
     H—C═C—C—NH2
        |       ‖
     H—C═C—C—H
            |
           NO2
```

2-Amino-4-nitropropoxybenzene

—NH_2 group in the original molecule by another NO_2 group forms 2-4-dinitropropoxybenzene, which is not sweet, but bitter.

The compound, dulcin, has a very sweet taste. Its composition is:

```
              H   H
              |   |
          O—C—C—H
          |   |   |
          |   H   H
          |
     H—C═C—C—H
        |       ‖
     H—C═C—C—H
            |
       H—N—C—NH2
              ‖
              O
```

Dulcin

If, however, the oxygen atom in the

```
   |
  HN—C—NH2
      ‖
      O
```

group is replaced by a sulphur atom—i.e.

```
   |
  HN—C—NH2
      ‖
      S
```

the substance is no longer sweet at all, but bitter. Another example, both of the unexpectedness of the chemical configuration of very sweet substances and also of the small changes in composition which cause a major change in taste, is that of saccharine:

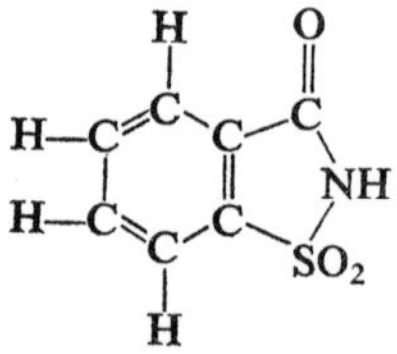

```
              H       O
              |       ||
          ,C.     ,C
H—C⁼       C—      \
    |       ||       NH
H—C⁼       C—       /
     C     ,C.     SO2
       |
       H
```

Saccharine

It is only necessary for the hydrogen atom of the $\rangle$NH group to be changed to a methyl group thus

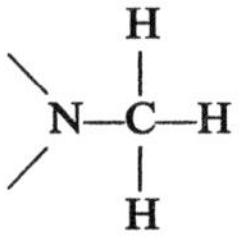

```
       H
 \     |
  N—C—H
 /     |
       H
```

for the sweetness to be entirely dissipated.

Moncrieff[4] has searched the chemical literature and drawn up a list of 58 chemical factors which he believes exercise a systematic effect on taste. The sourness of acids and the saltiness of ionisable salts of appropriate molecular weight have already been referred to. Other of Moncrieff's observations refer to the bitterness of iodides and of the salts of caesium and magnesium. Reference is also made to the sweetness of compounds containing amine groups and the bitterness of those containing two or three nitro groups.

The weakness of Moncrieff's attempted systemisation is that, in spite of a great deal of chemical research, the principles underlying chemical structure and taste are not yet entirely clear. Further, much of the existing codification can only be extended to quite simple tastes, such as sourness, sweetness and bitterness. Knowledge of more sophisticated tastes and the chemical nature of the substances that produce them, though fairly extensive, is still largely empirical. Table 15 shows the diverse chemical substances, the tastes of which are similar to fruits, such as pear, peach, pineapple, orange, apple, strawberry and banana; to onion, garlic and parsnip; to some of the spices; as well as to coconut and almond.

[4] Guenther, E. (ed.), *Kirk-Othmer Encyclopaedia of Chemical Technology*, vol. 9, 357. N.Y.

Table 15

The chemical structure of some synthetic flavouring agents

Flavour	*Synthetic flavouring agent*	*Formula*
Pear	n-Propyl acetate	$H_3C-CH_2-CH_2-O-C(=O)-CH_3$
Peach	γ-Undecalone	$H_3C-(CH_2)_5-CH-CH_2-CH_2-C=O$ (ring closed by $-O-$ between CH and $C=O$)
Pineapple	Methyl 3-methyl-thiopropionate	$H_3C-S-CH_2-CH_2-C(=O)-O-CH_3$
Orange	Decanol	$H_3C-(CH_2)_9-OH$
Lemon	Citral	$H_3C-C(CH_3)=CH-CH_2-CH_2-C(CH_3)=CH-CH=O$

Table 15 (continued)

Flavour	*Synthetic flavouring agent*	*Formula*
Apple	Isopentyl isovalerate	$(CH_3)_2CH-CH_2-CH_2-O-C(=O)-CH_2-CH(CH_3)_2$
Strawberry	Mixed ethyl acetate and n-pentyl acetate	$CH_3-CH_2-O-C(=O)-CH_3$ $CH_3-CH_2-CH_2-CH_2-CH_2-O-C(=O)-$ [cut off]
Banana	Methyl isovaleriate	$CH_3-O-C(=O)-CH_2-CH(CH_3)-CH_3$
Mustard	Ethyl isothiocyanate	$CH_3-CH_2-N=C=S$
Onion	Ethyl thiocyanate	$CH_3-CH_2-S-C\equiv N$
Garlic	Allyl disulphide	$CH_2=CH-CH_2-S-S-CH_2-CH=CH_2$

Table 15 (continued)

Flavour	*Synthetic flavouring agent*	*Formula*
Parsnip	n-Octyl butyrate and n-octyl propionate	H H H H H H H H H—C—C—C—C—C—C—C—C—H H H H H H H H	O H H H O—C—C—C—C—H H H H H H H H H H H H H—C—C—C—C—C—C—C—C—H H H H H H H H	O H H O—C—C—C—H H H H H
Pepper	Piperine	H H H H C——C C—C H H H H H H H H H—C—H N—C—C=C—C=C—C C——O H H O C=C—O—C—H C——C H H H H
Peppermint	Diethyl sulphite	H H H H H—C—C—O—S—O—C—C—H H H O H H
Vanilla	Vanillin	H H C—C O=C—C C—OH H C=C H H O—C—H H

Table 15 (continued)

<table>
<tr><th>Flavour</th><th>Synthetic flavouring agent</th><th>Formula</th></tr>
<tr><td>Coconut</td><td>γ-Nonalactone</td><td>H H H H H H H H
H—C—C—C—C—C—C—C—C—H
H H H H H | H |
O——C=O</td></tr>
<tr><td>Almond</td><td>Benzaldehyde</td><td>H H
C=C H
H—C C—C=O
C—C
H H</td></tr>
</table>

These are only a few of the chemical compounds, any one of which can be synthesised, which could be added to a mixture of synthetic protein, fat and carbohydrate to make the resulting combination, suitably enriched with synthetic vitamins, taste and smell something like one or other of the natural foods to which people have become accustomed and for which they have consequently developed a liking. In fact, although each of the compounds listed in Table 15 has a taste or smell reminiscent of the article recorded in the first column of the table, it is not identical with it.

I have already referred to the complexity of the sensory perception of taste in relation to the number and sensitivity of the taste buds. But the taste of any food-stuff is much more than the sensation recorded by the nervous recorders categorised as taste buds. The sensation also includes the far more subtle sensations of smell. H. H. Wright[5] has estimated that professional perfumers with highly trained noses can distinguish between very large categories of substances, probably in the millions. If it were assumed that there was only one kind of nerve receptor in the nose perceiving smell, then only two patterns of smell could exist, those capable of switching the

[5] Wright, H. H. (1968), *Sci. J.*, 57, July.

receptor on and those incapable of doing so. With two primary receptors, there are two possible patterns for the first and two for the second, that is, four in all. Similarly, with three receptors, 8 patterns of smell would be possible. Should there be twenty, however, then 1 048 576 categories of smell could be differentiated. And this is something of the order of the number of different smells there seem in actual fact to be.

Since this immense power of discrimination exists for smell, quite apart from the considerable, even if not quite so extensive, abilities to identify taste, it would be unreasonable to expect a single synthetic chemical to reproduce the complex sensations provided by the taste of whisky, roast pheasant or even pepper. It is for these reasons that natural pepper is still an article of trade of substantial economic importance. Similarly, even when flavours are prepared for use as an ingredient in manufactured foods, while they may contain one or more pure synthetic chemicals to contribute to the final effect, they almost always include as well concentrates or essences made from natural food-stuffs or flavouring agents.

Vanillin, whose molecular constitution is shown in Table 15, is the main component of the taste of vanilla. But although this is so, a simple solution of vanillin would never be confused with an extract of vanilla pods. Further, to someone of discrimination, the distinctive character of vanilla from Madagascar can be distinguished from vanilla from Mexico, and both possess flavours differing from that of Tahitian vanilla. That there is a real difference in the mixture of components present together with vanillin in the extracts from vanilla pods is shown in Figure 6. In this chart, every peak on the curves indicates the presence of a separate chemical compound. The identity of the compounds had not been established by H. W. Jackson,[6] by whom the graphs were published. Nevertheless, they indicate the complexity of the mixtures of which extracts from natural products are composed.

As I have pointed out in earlier chapters, food, besides being a vehicle for the nutrients by which life and health are maintained is also (like every other commodity, including houses and clothes and a decent standard of living, which includes quite sophisticated articles such as books, television sets and motor transport for visiting one's friends—all of which play a part in the maintenance of life and

[6] Jackson, H. W. (1966), *J. Gas Chromatog.* 4, 196.

happiness of which health is compounded) an item of economic significance. That is to say, people buy it because they want it. And they often want one particular food product rather than another because they like its taste. It is not surprising, therefore, that food scientists, even if they cannot exactly imitate the taste of natural foods, can at least attempt to equal them in attractiveness by something artificial and entirely new. After all, a Beethoven symphony is an artificial mixture of noises unknown to nature, but, in many people's opinion, *better* than any natural sound. The taste of Coca Cola is also unknown to nature.

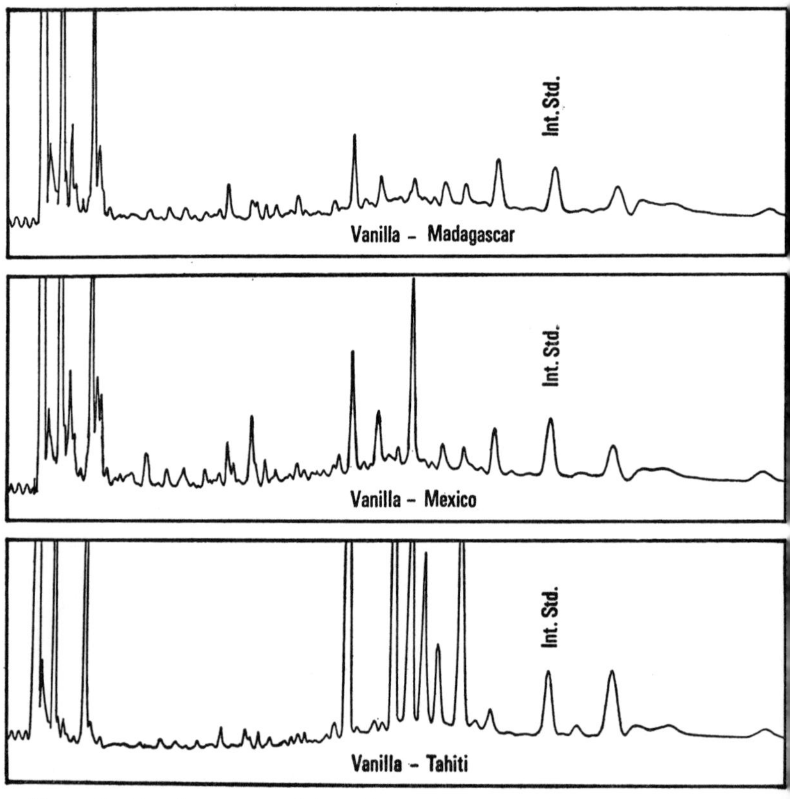

Figure 6 Gas-chromatographic analyses of extracts of three types of vanilla from which the vanillin has been separated.

In pursuit of high artistic merit in taste, food technologists have used and mingled a very large number of synthetic chemicals. The list given in Table 15 shows those which have been considered and approved, with certain provisos in some instances, by the United States public-health authorities. And this numerous list is only part of the armoury of a food technologist; in addition to such synthetic chemical compounds he can mingle items from an almost equal number of extracts and concentrates from natural sources.

The list of compounds shown in Table 15 comprises substances all of which have been considered by the United States authorities, not from the point of view of their taste but rather from their toxicological safety. Many of them which have been used traditionally for a long period of time are classified as 'generally recognised as safe' (GRAS). Others, however, have been tested and are accepted on the proviso of their being used in amounts not exceeding a particular level. The long list of several hundred compounds countenanced by the American authorities is in contrast to the list of only 11 synthetic substances, those 'not found in nature', to which the German official list is restricted.[7] The principle guiding German thinking is that foods should preferably be flavoured with substances derived from natural sources and that, even when synthetic compounds are employed, and even if they have been tested and found to be harmless, the articles containing them should bear a declaration stating that they contain an 'unnatural' flavouring agent.

Up-to-date analytical techniques have already shown that natural food-stuffs contain vast numbers of chemical constituents that together give them the taste and smell which combine to produce the character for which they are esteemed. There is therefore justification for the extensive list of some 840 chemical compounds set out in a recent article.[8] On the other hand, public-health authorities have cause for concern. Let me give two examples.

Glutamic acid is one of the several amino-acids of which food proteins are composed. The sodium salt of glutamic acid, monosodium glutamate, contributes a meaty flavour if it is added as a condiment to foods and, in addition, is claimed to accentuate other flavours already present in the food, in the same way that pepper is claimed to 'bring out' the flavour of strawberries. Monosodium glutamate (MSG) may be manufactured from protein, produced by

[7] Vodor, C. A. (1964), *Manufac. Chem. & Aerosol News.*, Sept.
[8] Hall, R. L. (1960), *Food Tech.*, 14, 491.

microbiological fermentation or synthesised. There are certain other substances, notably 5′-inosinate and 5′-guanylate which also act in the same way as MSG as flavour 'potentiators'.[9]

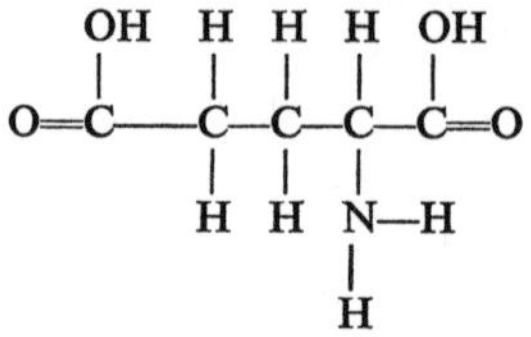

```
      OH    H   H   H   OH
      |     |   |   |   |
  O==C-----C---C---C---C==O
            |   |   |
            H   H   N—H
                    |
                    H
```

Glutamic acid

It could safely be assumed, one would imagine, that a simple amino-acid which occurs naturally in most common food proteins, could be added to food without harmful results if, by its presence, it improved the flavour. Yet in April 1968, a communication from a Dr Ho Man Kwok was published in the *New England Journal of Medicine*[10] describing symptoms, including a feeling of numbness in the nape of the neck gradually extending to the arms and back together with 'general weakness and palpitation'. Quite soon other reports of similar effects appeared.[11] All of these related the symptoms to the consumption of Chinese food, and it was soon discovered that the cause was heavy seasoning of the food with monosodium glutamate.

This distressing but evanescent 'poisoning' with a specific flavouring agent is perhaps a trivial example of what it is feared may happen unless reasonable precautions are taken. A more subtle and possibly more potentially serious example is that of the group of compounds called the cyclamates. Cyclamic acid, whose precise systematic name is cyclohexanesulphamic acid, possesses the chemical structure shown opposite.

Cyclamic acid and its sodium and calcium salts are synthetic sweetening agents. They are about 30 times as sweet as sugar, compared with saccharine which is 350 times as sweet. Cyclamates, however, possess a somewhat more agreeable taste and, since they

[9] Kuninaka, A. (1964), Kibi, M., and Sakaguchi, K., *Food Tech.*, 18, 29.
[10] Kwok, R. H. M. (1968), *New Eng. J. Med.*, 278, 796.
[11] Schaumburg, H., *ibid.*, 278, 1122; McCaghren, T. J., Menken, M., Mugden, W., Rose, E. K., Rath, L., Beron, E. L., Kandall, S. R., Gordon, M. E., Davies, N. E., *ibid.*, 278, 1123, 1124.

```
        H       H
        |       |
        C———————C   H H
       /|       |\  | |
  H\  / H       H \ | |
    C              C—N—SO3H
  H/  \ H       H /
       \|       |/
        C———————C
        |       |
        H       H
```

Cyclamic acid

share with saccharine the property, emphatically not possessed by sugar, of possessing no nutritional value at all—in particular, they do not make people fat or encourage coronary heart disease—they have since 1950 been added in increasing amounts to soft drinks and manufactured foods.

During the nine years from 1941 to 1950, the safety of cyclamates was exhaustively tested by feeding trials, first on animals and then on man. Judged by these tests they appeared to be harmless in the amounts that it could be anticipated that they would be consumed.[12] The public-health authorities in both the United States and Great Britain consequently approved their use. And very popular they became. In seven years, the amount of cyclamates consumed in the United States increased twentyfold. By 1965, 5000 tons were eaten and drunk. From being examples of a synthetic flavour, cyclamates became promoted to the status of the synthetic non-foods, which I discuss in more detail in the next chapter. They became popular ingredients of 'special purpose' foods and 'low-calorie' drinks. As sugar substitutes they took their place in articles specially designed for slimming. These covered powdered drink mixes, syrups, canned fruit, jam, jelly, preserves, desserts, salad dressing and baked goods.

In consequence of their popularity, the total amount of cyclamate eaten by single individuals rose to quite substantial proportions and the public-health people began to think that their previous trials, which had been conducted on the basis that not much of a flavouring would be consumed by one person, should be repeated. The sugar manufacturers, whose commodity, as I have already pointed out, far from becoming scarce as the world population increases, tends to come on the market in surplus amount, also considered that if cyclamates were indeed toxic there was no harm in letting the public

[12] Anon. (1966), *Lancet.*, *i*, 134.

know. The sugar interests consequently financed the Wisconsin Alumni Research Foundation[13] to carry out tests of the effect of feeding rats on diets containing up to 10% of their weight of cyclamate. In fact, the tests, beyond showing some retardation in growth of the second generation of rats fed throughout their own lives and those of their parents on diets containing the higher levels of cyclamate sweetness, provided no conclusive demonstration that cyclamates were harmful. The incident demonstrates, however, how difficult it is to prove a negative, namely that a new substance does *not* possess toxicity. Later work led to cyclamates being abandoned (see *Lancet* 1970, i, 1091.)

A second, equally subtle, example of the same thing is that of coumarin. Coumarin, whose chemical configuration is shown below, is a natural component of many foods. It occurs, for example, in strawberries, and contributes a part of the agreeable and characteristic

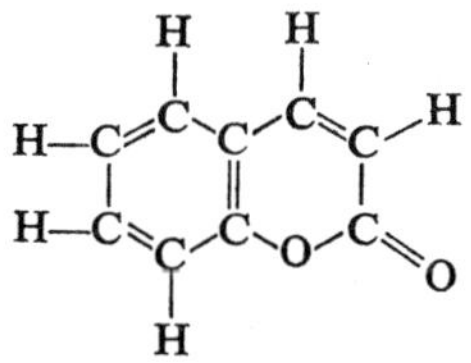

Coumarin

flavour of these fruits. It is comparatively easy to synthesise, and was in fact first prepared as long ago as 1875 by W. H. Perkin, the elder. Coumarin has an agreeable taste and smell somewhat reminiscent of fresh hay. It was, therefore, not unreasonable to use it as a synthetic flavouring agent.

When coumarin is administered in large amounts, far exceeding the quantities in which its taste would be tolerable, it is known to be associated with an increased tendency to spontaneous bleeding. This property causes it to be used by physicians as a drug in the treatment of patients who have suffered a coronary thrombosis and for whom any tendency to blood clotting could be particularly dangerous. There are, therefore, several thousand apparently normal people who are receiving carefully adjusted doses of coumarin. Although coumarin used as a flavouring agent is innocuous to the great bulk

[13] Nees, D. O. (1965), and Dorse, P. H., *Nature, Lond.*, 208, 51.

of people who might eat foods flavoured with it, for this quite substantial minority its presence could constitute a danger.[14]

In spite of the possibility of toxicological danger, in spite of the inevitable need for careful and rigorous testing to ensure the demonstration of what is a philosophically unattainable negative virtue, namely, absolute safety, the organic chemist, as is demonstrated by Table 15, can provide a long list of synthetic flavours. But although a rich diversity of such synthetic flavouring agents can be manufactured, the resulting tastes, interesting and attractive though they may be, cannot so far be made identical with those derived from nature. This may not be important; it is nevertheless worth recording.

Colour, like taste, is an important attribute of the attractiveness of food. Unlike the imitation of natural tastes, however, the reproduction of natural food colours by artificial means does not present the chemist with any particular difficulty. The initial achievements of Victorian organic chemists in establishing a dye-stuffs industry based on coal-tar chemicals has been developed and extended during the last 100 years or so. This has led to the elaboration of an extensive group of dye-stuffs by means of which any desired shade and hue can be obtained. In recent years, the public-health authorities, as a result of their own tests, the philosophy of public-health policy in their own country, as well as the general advance in scientific knowledge, have in each country selected a slightly different list of synthetic colours that can permissibly be incorporated in food-stuffs. Table 16 shows the permitted lists of several countries in 1970.

The reason for the lack of unanimity of authorities in different countries arises from two causes. The first is that, in the main, the synthetic dye-stuffs used to colour foods belong to the group of compounds commonly classified as 'coal-tar dyes' because they were mainly derived first of all from this source. The second is that the compound, 3:4-benzpyrene, which possesses a structure similar to that of many other substances of the same group, is known to be carcinogenic. Benzpyrene is, to be sure, different from the dye-stuffs we are discussing; nevertheless this may not be a guarantee of safety. For many years the oil-soluble dye, *p*-dimethylaminoazobenzene, so-called Butter Yellow, was included in the United States list of approved food colours. In 1932, however, a Japanese worker, M.

[14] Kekwick, A. (1966), *J. Inst. Food Sci. Tech.*, 1, 1.

Table 16

Synthetic food colours permitted in several different countries

Dyes	*Great Britain*	*Australia*	*Canada*	*Denmark*	*Finland*	*Germany (W)*	*India*	*Norway*	*Spain*	*Sweden*	*Switzerland*	*South Africa*	*USA*
Ponceau 4R	+	+		+	+	+	+	+	+	+	+		
Carmoisine	+	+		+	+	+	+	+	+	+	+	+	
Amaranth	+	+	+	+	+	+	+	+	+	+	+	+	+
Red 10B	+			+									
Erythrosine BS	+	+	+	+	+	+*	+	+	+	+	+*	+	+
Red 2G	+			+								+	
Red 6B	+			+			+						
Fast red E	+	+		+	+		+	+	+	+	+		
Red FB	+	+					+						
Orange G	+			+								+	
Orange RN	+			+								+	
Oil yellow GG	+											+	
Tartrazine	+	+	+	+	+	+	+	+	+	+	+	+	+
Yellow 2G	+	+		+									
Sunset yellow FCF	+	+	+	+	+	+	+	+	+	+	+	+	+
Oil yellow XP	+		+										
Green S	+	+		+								+	
Indigo carmine	+	+	+	+	+	+	+	+	+	+	+	+	+
Violet BNP	+	+		+									
Brown FK	+	+		+									
Chocolate brown FB	+	+		+									
Chocolate brown HT	+	+		+									
Black PN	+	+		+	+	+	+	+		+	+		
Black 7984	+					+							

* For colouring whole, halved or stoned fruit only.

Yoshida,[15] showed it to be capable of producing liver tumours in rats. The difficulty—nay, the impossibility—of prognosticating from its chemical structure whether or not a dye-stuff is likely to possess carcinogenic activity can be seen from Table 17, in which, besides benzpyrene, are shown the chemical structures of two dyes which are known to cause cancer, and two others which are used to colour foods.

[15] Yoshida, M. (1932), *Proc. Imp. Acad. Japan*, 8, 464.

P. R. Peacock, when he was director of research of the Glasgow Royal Cancer Hospital, pointing out[16] the difficulty of separating harmful from harmless dyes, drew attention to the fact that as long ago as 1895 cancer of the bladder had been diagnosed as causing the

Table 17

The chemical structure of synthetic dyes now known to be carcinogenic, and of others used to colour foods

1. *Carcinogenic*

[structural formula]	*3:4-Benzpyrene*
[structural formula]	*Sudan I*
[structural formula]	*Butter Yellow*

[16] Peacock, P. R. (1952), *Chem. & Ind.*, 238.

Table 17 (continued)

2. *Used as food colours*

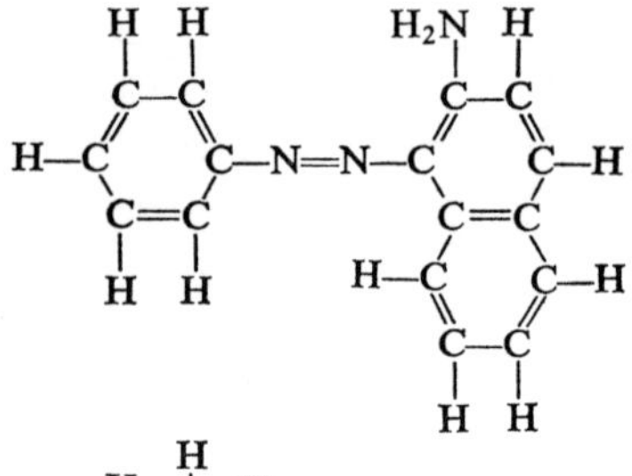

Yellow AB

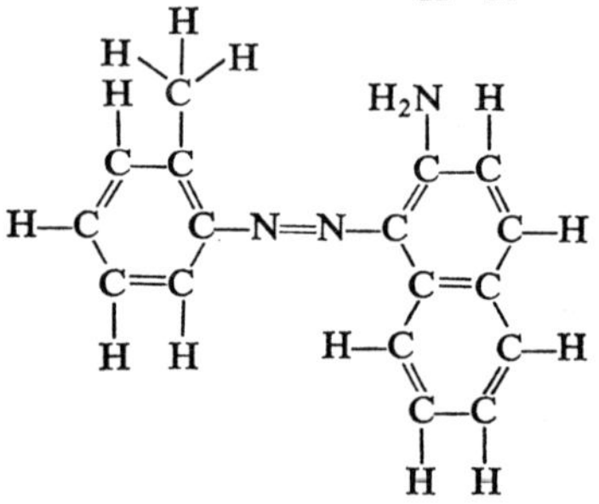

Yellow OB

death of workmen handling coal-tar dyes and that one of the compounds incriminated in causing it was β-naphthylamine, whose structure is shown below:

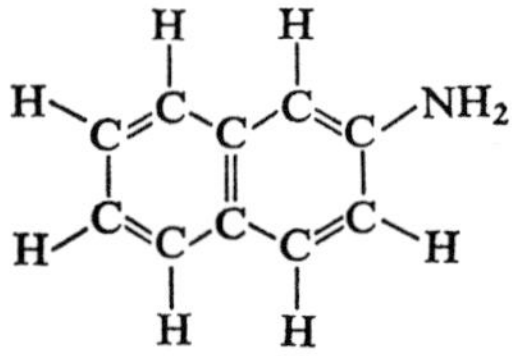

β-Naphthylamine

Perhaps most telling of all, Peacock predicted that 20 years or more might elapse before it could definitely be proved that a particular chemical compound had been the cause of cancer from which people a generation later were dying.

The conclusion can, therefore, be drawn that, although it is not difficult to synthesise chemical dyes to colour any type of food, it may be prudent for the organic chemist not to do so. Table 16 showed the list of synthetic colours which are permitted to be used to colour food. Table 18 shows a further list of pigments derived from biological sources which effectively colour the commodities to which they may be added, but which may be less strange—and consequently less potentially harmful—to the tissues of the body of the person eating them.

Table 18

Colouring matter derived from biological sources used to colour foods

Red:	Alkanet
	Beet, Betanin
	Cochineal
	Carminic acid
	Sandalwood
Orange:	Brazilwood
Yellow:	Annato
	Bixin
	Carotene
	Persian berries
	Saffron
	Crocin
	Turmeric
	Curcumin
	Xanthophyll
Green:	Chlorophyll
Violet:	Orchil
Blue:	Anthocyanin

Both flavour and colour are important attributes of foods. Many unusual food products, often possessing good nutritional value—dried yeast, leaf protein, algae, fish flour and many more—have not fully succeeded in their purpose because their flavour, colour and texture were unacceptable. The synthesis of flavour and colour is, therefore, as potentially useful as the synthesis of protein or fat. So far, however, it seems that the production by synthesis of a flavour the same as that of natural food-stuffs has defeated the ingenuity and knowledge of the chemist. This is because of the complexity of the mixtures of components, some of which are present in extremely minute amounts, of which natural flavours are composed. Synthetic foods are, therefore, likely to taste different from natural ones; they may, however, be none the worse for that.

Natural food colours, on the other hand, can readily be simulated by synthetic dyes. For the present, nevertheless, the use of such dyes may diminish, or may be discontinued altogether, because of the difficulty of being certain that, in the long run, they may not be shown to be carcinogenic or otherwise toxic.

CHAPTER 8

TEXTURISATION

Synthetic vitamins and amino-acids are already manufactured on an industrial scale and used as part of both human and animal diets. Fat has been synthesised as a major commodity and knowledge is available to show that protein and carbohydrate could be produced synthetically. Flavours and colours, too, can be made in a chemical plant to enliven any mixture of synthetic nutrients which may be compounded. But in spite of all these scientific advances, there is still more to be done. Foods need to possess acceptable structure and consistency as well as an appropriate chemical composition and the right smell and taste.

One of the major preoccupations of the commercial baker, and of the food scientists who support him, is how to obtain what is described as a 'bold, well-set-up loaf', with uniform crumb structure, springy in consistency and silky in texture. To achieve this target, fundamental studies were carried out of the chemistry of flour protein and of its reaction when submitted to varying degrees of stress and working. The stretching and kneeding of dough, at one time done by hand, now carried out by machinery, exerts a similar function on the molecular organisation of the gluten fibres in it as does the 'cold drawing' of nylon. The theoretical basis upon which the discovery of flour 'improvers' is based is similar to that upon which similar chemical agents used to influence the structure of the protein, keratin, in hair in solutions designed to produce 'home perms' depend. A chain of research led, first, to the discovery of nitrogen trichloride, called 'agene' by the baking industry, which as an 'improver' was widely used in both Europe and the United States for a generation or more, and then to the discovery that it could produce in bread a toxic substance, methionine sulphoximine. It culminated in the abandonment of nitrogen trichloride and its substitution by, among other agents, chlorine dioxide. The researches involved were all undertaken solely to bring about a small but important change in loaf size and crumb structure. In other words, all this scientific effort and toxicological research were carried out to attain what was accepted as being the appropriate structure for a loaf of bread.

Problems of the physical structure and consistency of foods must be accepted as important, just as are those of colour, taste and smell. While the physiological efficiency of the body tissues depends on the satisfactory chemical composition of the diet in terms of the nutrients it contains, the well-being of the individual person within the social structure of the community of which he is a member depends on much else. I have written about this elsewhere.[1] Although questions of structure and consistency do not apply particularly to synthetic food, but are equally relevant to natural foods as well, they cannot be overlooked in a discussion of the possibilities of food synthesis. The purpose of the scientist who sets out to manufacture synthetic food is to produce it at a price within the economic reach of the community for whom it is intended; it is not to change the behaviour of the community. It is, therefore, sensible for such a man to aim either to produce components which can be used as ingredients of accepted food commodities, or to present his synthetic articles with a taste, appearance and consistency similar to those of foods which already fit into the pattern of behaviour of the society for which they are intended. There was, for example, no technical reason why the German manufacturers of synthetic fat should not have marketed it as a free-flowing oil or, alternatively, as stiff as a block of suet. It was sociologically sensible of them to make it as similar in consistency to butter as they could.

Butter, and the margarine which is its technological equivalent, are, for the most part, designed to be spread on bread. They are better liked and, in consequence, considered to be of higher quality, if they can be spread without tearing the bread to pieces. Considerable scientific effort has been devoted to attempts to measure the 'spreadability' of butters and margarines. For example, measuring instruments, some of considerable ingenuity and complexity, have been developed to imitate as closely as possible the action of spreading butter on bread. The operation could typically involve the shearing of a small cube of the fat under test by a knife-edge moving at a series of different speeds and pressing the fat down on to the roughened surface of the piece of artificial 'bread' with varying degrees of

[1] Pyke, M. (1968), *Food and Society*. John Murray, London.

force.[2] The complexity of the physical principles controlling even so apparently simple a property as this is, however, shown by the fact that the results obtained by these instruments specially designed to give precise quantitative results did not agree particularly well either with measurements of viscosity, flow, resistance to shearing and deformation to stress recorded under strict laboratory conditions or to the opinions about the performance of the same fats expressed by a panel of judges.

During the manufacture of butter and margarine, close attention is paid to the precise way in which the water is incorporated with the fat, the presence of exact amounts of salt, emulsifiers and other ingredients capable of affecting the physicochemical equilibrium and, most important of all, the degree of work imposed on the mixture at an exactly controlled temperature by the 'votator' or other similar machine through which the fat is passed during the manufacturing processes. It is now apparent that the success or otherwise of all these operations is reflected in the precise structure and arrangement of the fat crystals in the final product.[3]

All this effort and expense and the considerable investment of research and scholarship are not made for fun, but to insure that the structure of the final product shall be what the people who eat butter or margarine think that it ought to be. This is not necessarily a frivolous objective; indeed, it may be a necessary one. The roots of human behaviour, and particularly behaviour in eating, rise from deep strata.

The problem of presenting synthetic protein as a food-stuff or food ingredient with some sort of recognisable and acceptable structure and appearance may well be more difficult than that arising from synthetic fat. Fats must possess acceptable rheological characteristics, but, although their crystal structure is important, they have no recognisable structural anatomy or 'grain'. Many protein foods, and particularly meat, do. Although beef steak ought not to be tough, neither should it possess the softness of marshmallow. The structure of meat, which gives it one of the culinary qualities for which it is esteemed, is due to the orderly arrangement of muscle fibres of which it is composed. If synthetic protein is to

[2] Mohr, W. (1949), and Hassing, J., *Milchwiss.*, 4, 255; Coulter, S. T. (1936), and Combs, W. B., *Minn. Agric. Exp. Stn. Tech. Bull.*, 115; Prentice, J. H. (1952), *Brit. Food Manf. Ind. Res. Assoc. Rep.*, 37; Prentice, J. H. (1956), *ibid.*, 69.

[3] Gunstone, F. D. (1964), *Chem. & Ind.*, 84, 18 (Jan.).

compete with meat, not only as a contribution to the nutritional requirements of the body, but also to the attractiveness of a good meal, it will need to be fabricated into something not entirely dissimilar to meat. Synthetic protein has not, so far, been prepared on a sufficiently large scale to allow its being submitted to trial. Attempts have, however, been made to solve the similar problem of how to present protein isolated from readily available vegetable sources as if it were meat. The process on which most work has been done has been that in which the protein is made up into a viscous solution, extruded through small holes as an elastic thread, wound into a hank, mixed with other ingredients and flavours, and then cut into slices across the grain.

Variants of this process have been tried.[4,5] By changing the concentration of the protein solution, the composition of the precipitating bath and the mechanical treatment of the fibres after they are produced, their character can be varied from that of delicate cobweb suitable for conversion into tender 'chicken' to tough strands capable of simulating stringy mutton. The process described by Thulin and Kuramoto[6] for making a simulated meat out of soya-bean protein is thus. Disperse the purified protein in an alkaline solution and extrude it through the same sort of spinnerettes as are used in making nylon fibres. The exuded protein is allowed to pass into a coagulating bath, which is usually composed of a solution of acid and inorganic salts. The diameter of the filaments may be varied from 1 to 30 thousandths of an inch. Once these fibres of protein are formed, they are blended with fat, flavours and colours may be added, vitamins incorporated and perhaps a preservative as well. The fibres, wound into a hank or skein, are then bound together and made into any desired form—as slices or cubes or into granules like minced meat. The objective of the whole exercise is to produce something with the sort of structure that people would want to eat.

In 1958, G. W. Scott Blair,[7] who had made a life's work of the topic, published an extended review of the theoretical basis, the technological methods available, and the significance to the consumer of rheology in food science. For every type of food it is quite

[4] Lundgren, H. P. (1949), *Adv. Protein Chem.*, 5, 303.

[5] Odell, A. D. (1967), *Proc. Conf. Soybean Protein Foods*, U.S. Dep. Agric., A.R.S.-71-35, 163.

[6] Thulin, W. W. (1967), and Kuramoto, S., *Food Tech.*, 21, 168.

[7] Scott Blair, G. W. (1958), *Adv. Food Res.* 8, 1.

clear that its structure and consistency are matters of importance. In general terms, the food technologist refers to these problems as 'texturisation'. The instruments for assessing the physical characteristics involved may measure fundamental properties, such as, for example, crystal structure. Or they may only record empirical properties in scientific terms, such as the 'alveograph' which draws a graph to show the strength of a bubble of dough, thus giving a measure of how a particular sample of flour will behave, or the 'tenderometer' which—as could be imagined—measures the tenderness of green peas from the field and hence shows when they should be harvested.

Food technologists manufacturing bread, cheese, margarine and butter, chocolate, frozen peas, jellies, breakfast cereals—the list could be extended throughout the gamut of foods—all need to study the structure of their product. The same problem, in a heightened form, will need to be faced by those who synthesise food, because the material they make will, by its very nature, lack any recognisable structure of its own.

CHAPTER 9

SYNTHETIC NON-FOODS

For most of the people in the world most of the time eating is in real earnest. But for all people, eating some of the time is a pleasure rather than a necessity. I have written elsewhere[1] of the various motives and the diverse circumstances involved in the consumption of particular articles of diet under peculiar circumstances. And in the industrialised countries of the West, in Japan, South Africa, Australia and elsewhere in a steadily growing area of the earth, the same scientific and technological expertise which has made possible the synthesis of nourishing food, carefully 'texturised', coloured and flavoured synthetically has at the same time brought wealth. This wealth provides machines to do work otherwise done by human muscle power, for the execution of which muscular endeavour it was necessary to eat. But at the same time that the need for physical effort has been reduced, growing wealth has made it possible for the possessors of the wealth to buy more and more various, sophisticated and elaborate food. As a result, diseases of malnutrition due to not having enough to eat or to not having enough protein, vitamin A, B vitamins or other nutrients have decreased in importance, and another kind of malnutrition has appeared instead. This has been obesity, a disease which shortens life and encourages diabetes, coronary heart disease and varicose veins, quite apart from causing distress, particularly among Western women, who believe that it detracts from their good looks.

It follows from this that the chemical synthesis of food can be useful, not only as a source of edible commodities to provide nourishment and thereby supplement the exiguous supplies of natural foodstuffs, but also to provide articles which can be eaten with pleasure and interest, but which will, nevertheless, do the consumer no good whatever and, most of all, which will not make him or her fat.

A beginning has already been made in the production of artificial food which, while being attractive to eat, will be without value as food. A gram of sugar provides the body with 4 calories, a teaspoonful

[1] Pyke, M. (1968), *Food and Society*. John Murray, London.

with 20. Anyone who takes, say, three teaspoonfuls of sugar in each cup of tea and drinks two cups for breakfast, another mid-morning, a fourth mid-afternoon and two more before bed-time obtains 360 calories from the sugar in them without noticing having done so. Sugar is in two ways the most insidious of all carbohydrates. Its sweetness, which appears of itself to be of no physiological significance to man, regardless of its attraction for bees and its consequent survival values to the plants whose flowers the bees pollinate, is pleasant to many people's taste. In addition, sugar, because of its solubility in water, does not make the foods in which it is incorporated any bulkier than they were before. There is, therefore, no direct indication that the sweetened article is contributing to the body of the man or woman eating it more food value, in terms of calories, than the same article unsweetened. People who like sweet things consequently tend to eat more than they need, and become fat. The synthetic chemical, saccharine, is, therefore, a non-food substitute for sugar, even though it contributes only the taste of sweetness. On the other hand, cyclamates, which were discussed in Chapter 7, serve as a substitute both for the sweetness of sugar and, because they are used in amounts which, while less than the amounts in which sugar would be employed, are at least of a similar order of magnitude, for the 'body' of sugar as well. This is one reason for their popularity in soft drinks designed to provide very few calories, or none at all. And, as I mentioned in Chapter 7, this very popularity, which has led to increasing amounts of cyclamates being consumed, is the direct cause of the concern of public-health authorities at the possibility that these increasing quantities may exert a harmful effect on those consuming them, even though cyclamates possess no nutritive value.

This highlights the basic problem of devising a non-food. It must either possess a molecular size so large that it is not absorbed into the system or, if it is absorbed into the blood-stream, it must pass out again through the kidneys without exerting any direct physiological effect. The cyclamates are strongly ionised and are obviously of the appropriate molecular shape and size to affect the taste buds with a sensation of sweetness. At the same time, very little or any of them appears to be absorbed. There are, however, many more groups of compounds, less physiologically unusual than the cyclamates, from which non-nutritional non-foods could be made. Some of these occur, as it were, unintentionally in foods; others are already used for their lack of nourishment.

The plants we eat as food under the title of 'vegetables', including some fruits, often contain, as well as starch and sugar, at least two other groups of compounds: hemicelluloses and fibre. In young plants the whole of the cell wall is composed of cellulose. This, as I have already described in Chapter 5, is a polysaccharide made up of glucose molecules, but because of the way in which the glucose units are linked together, it is not susceptible to human digestive enzymes and can, therefore, be categorised as a non-food. Herbivora are the higher animals best capable of digesting cellulose; this they do with the assistance of the micro-organisms living in their guts. Curiously enough, German scientists[2] have found that man is little, if at all, inferior to the cow in this respect, particularly if the cellulose is fed in a finely divided state. However, although man may be capable of utilising small amounts of cellulose he needs enough time to do it and quantities sufficiently large to make a significant contribution to his diet do not remain long enough in his digestive tract. McCance and Lawrence[3] calculated that the maximum nourishment attainable by an average man from cellulose would not exceed 18 calories in a 24-h day.

It is only in the young plant that the cell wall is composed solely of the polysaccharide, cellulose. As the plant grows, other compounds are synthesised. Among these is lignin, which is one of the structural polymers of wood and which, therefore, contributes to the stiffness and strength of the plant's structure. But although lignin itself, like cellulose, is indigestible and is from its nature inappropriate as a component of an artificial food-stuff—after all, it would be naïve to promote sawdust as a food-substitute, even though it is non-fattening—several intermediate compounds are to be found in vegetables in which the synthesis of lignin is occurring. These might well be useful as presumably innocuous components of non-foods. Curiously enough, the exact structure of the polymer of which wood lignin is composed is not fully understood. It is, however, known to be made up of units of the three cinnamyl alcohols:

[2] König, H. A. (1902), and Reinhart, E., *Z. Nahr. Genussm.*, 5, 10; Strauch, W. (1913), *Z. Exp. Path. Tier.*, 14, 462; Rubner, M. (1916), *Arch. Anat. Physiol.*, 37.

[3] McCance, R. A. (1929), and Lawrence, R. D., *Med. Res. Coun., Spec. Rep. Ser.*, 135.

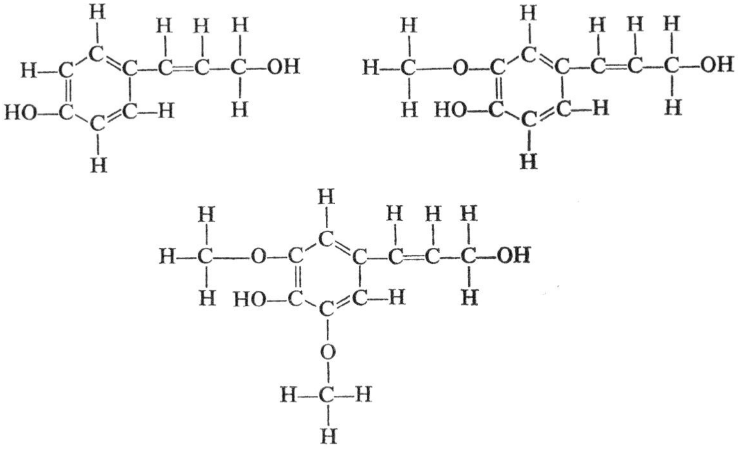

```
        H                                      H
        |    H  H  H           H               |     H  H  H
        C    |  |  |           |               C     |  |  |
 H—C⁄     ⧵C—C=C—C—OH      H—C——O—C⁄     ⧵C—C=C—C—OH
   ||       |       |          |       ||      |        |
 HO—C⧵   ⁄C—H       H          H  HO—C⧵   ⁄C—H        H
      C                                   C
      |                                   |
      H                                   H

                        H
              H         |     H  H  H
              |         C     |  |  |
          H—C——O—C⁄     ⧵C—C=C—C—OH
              |      ||      |        |
              H  HO—C⧵   ⁄C—H         H
                        C
                        |
                        O
                        |
                      H—C—H
                        |
                        H
```

Cinnamyl alcohols

Although the lignin in fully formed wood is obviously too chemically stable and mechanically rigid to eat, the partly formed polymer could well provide a non-digestible and non-absorbable material which might yet, when suitably prepared and flavoured, be eatable. It is perhaps worth noting that certain of the polymer units, chemically broken off the oak of wooden casks by the action of alcohol, are in fact consumed by those who drink spiritous liquors which have been aged 'in the wood'.[4]

As plants grow older, the proportion of lignin in their cells tends to rise. Woodiness in the thicker cabbage stalks, for example, is due to lignin formation. So far, lignin has not been used as a non-food. There are, however, at least two analogous groups of compounds which have been employed in foods, and in the use of both of them a considerable degree of organic chemical knowledge has been used to fit them for this purpose. Intermediate between the simple sugars which are synthesised in the leaves of plants and the large polymeric structure of the cellulose molecules come groups of intermediary compounds, often loosely categorised as 'hemicellulose'. Among these are the pectins. Pectins are extensively used in manufactured foods, not on account of their nutritive value (indeed, they are probably unavailable to the body—hence their potential usefulness in

[4] Duncan, R. E. B. (1966), and Philp, J. M., *J. Sci. Food Agric.*, **17**, 208; Sisakyan, N. M. (1951), *Dokl. Akad. Nauk.*, **79**, 639.

intentionally non-nourishing articles), but because of their technical properties of forming jelly.

It will be recalled from Chapter 5, that whereas starch, which is a polymer of glucose units, is digestible, and after digestion serves as the main source of metabolic energy, cellulose, which is also a polymer of glucose units, is not digestible and only provides a minimum of utilisable energy. The reason for the non-food character of cellulose is that the glucose units are linked, as it were, alternatively 'up' and 'down', by the so-called β linkage, whereas in starch the glucose units are connected by α linkages and are, so to speak, all the same way up. I reproduce the structure of cellulose below:

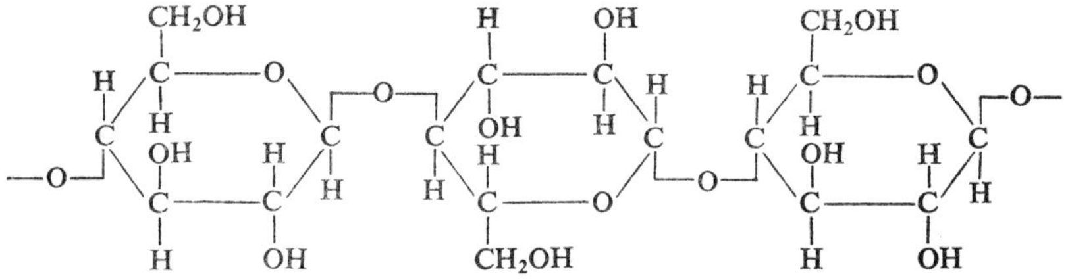

Cellulose fragment

Pectin, like cellulose, is mainly composed of a chain of sugar molecules connected by β linkages, and consequently, like cellulose, it is not digestible and therefore makes no contribution to nutrition. But, unlike cellulose, because some of the sugar units have attached to them methyl groups, $—CH_3$, and possess the configuration of galacturonic acid, which is not strictly a sugar at all, pectin, although a non-food, is eminently eatable. It is, in fact, the substance that causes jam to gel. The basic structure of pectin is shown below:

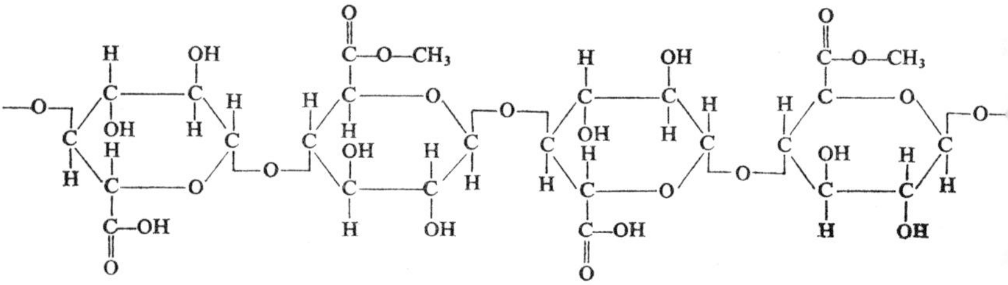

Pectin fragment

Surrounding the chain structure of the pectin molecule are various sugars, among which arabinose, galactose and rhamnose are prominent. The ways in which these are linked chemically to the main chain have not been fully elucidated. It is the variants in structure and the length and configuration of the chain that give the variation in properties and performance found to exist between different pectin preparations.

At the present time, pectin is not synthesised but is extracted from such materials as the apple residues remaining after the preparation of cider or apple juice, from sugar-beet residues after the separation of the sugar, or from the peel of citrus fruits remaining after the extraction of their juice. The pectin is normally prepared by heating the dried apple pomace or other raw material with dilute acid, and subsequently precipitating the pectin by adding salts of aluminium or copper. The pectin is then filtered out and washed. But although pectins are not synthesised, their chemistry can be purposely modified to produce a material with a variety of desired properties. The higher the molecular weight, that is to say, the longer the chain, the more pronounced are the gel-forming powers. The more methoxy groups, —$O.CH_3$, there are in the molecule, the quicker the gel sets. On the other hand, pectins containing specially reduced numbers of methoxy groups in their molecules have been prepared,[5] with which solid gels containing little more than 1% of actual dry substance can be made. Here indeed, it would seem, is the basis of the almost perfect non-food.

Pectins and their derivatives have been used for the manufacture of jams and jelly, as a stabiliser in ice-cream and as a thickening agent in custard. But once their properties as a non-food are made the basis of serious study, their merits in this direction cannot fail to be recognised. Unlike cyclamate, the basic molecular units of pectin have been made familiar to the body in the plant foods in which they have been up till now a somewhat neglected component. Yet already, for the special technical uses to which it is put by food manufacturers, pectin is prepared as a commercial commodity in substantial quantities. For example, W. A. Bender, writing in 1959,[6] estimated that the probable annual production of purified pectin in the United States

[5] Baker, G. L. (1948), *Adv. Food Res.*, 1, 395.

[6] Whistler, R. L. (1959), and BeMiller, J. N. (eds.), *Industrial Gums*. Academic Press, N.Y.

and Europe together amounted to 6000–8000 tons. Whether or not pectins modified to provide the degree of hardness or softness of texture or consistency required to make artificial foods are ever synthesised completely or merely de-esterified or split by chemical means to an appropriate degree, their potential usefulness as the basis of non-food is unquestionable.

Another potential component of non-food can perhaps best be introduced by a quotation. 'A patent covering the production of artificial fruits and vegetables', wrote the journal, *Food Manufacture* in October, 1968, 'has been issued to General Foods Corporation, White Plains, New York. The patent covers a substance referred to as an edible, crisp, chewable, non-uniform agglomerate of calcium alginate cells with the addition of artificial flavourings and other substances. The cellular structure is produced by adding calcium salts to a water solution of alginic acid ... to form a gel structure. The purpose of the invention is described as a method of preparation of artificial fruits and vegetables which may be heated or cooked without losing their characteristic crisp texture'. A further property of these artificial fruits and vegetables, besides their resistence to cooking, would be their almost total lack of food value.

Alginates are at the present time manufactured in substantial amounts from several different species of seaweed. In the United States, *Macrocystes purifera* and *Laminaria* are mainly used, whereas in Europe *Laminaria* and *Ascophyllum* are employed. Like starch, cellulose and pectin, alginates are polymers. Again, like pectin, they are similar to starch, but, once more like pectin, the units of which the molecular chains are composed are not glucose but mainly mannuronic acid with some glucuronic acid. Once more, the chain linkage is unstarch-like, so that the alginates represent a further group of compounds which, it seems, may be classified as non-foods.

The chemical properties of alginates endow them with several technological virtues. Advantage has been taken of these virtues by food manufacturers to use alginates and their chemical derivatives for useful purposes. For example, the sodium salt of alginic acid is soluble, whereas the calcium salt is not. If, therefore, a dry mixture of sodium alginate and calcium phosphate is stirred into water, a quick-setting jelly is produced. The molecular configuration of the basic polymer chain of which alginic acid is composed is shown opposite.

Alginic acid fragment

Gels produced by the introduction of a calcium salt—calcium phosphate, calcium citrate, calcium carbonate, calcium sulphate and calcium tartrate have all been used, with slightly differing results—to sodium alginate causes the formation of a chemically set, irreversible solid that does not melt when it is heated. It has, therefore, been eagerly seized upon by food manufacturers for the preparation of puddings, desserts, aspic, meringue and icing, all of which can, if desired, be absolutely devoid of nourishment. For example, a milk pudding has been patented[7] based on a specially treated blend of water-soluble sodium alginate, a mild alkali, such as sodium carbonate, and a small quantity of calcium salt, such as calcium lactate. These components are not simply mixed together but are prepared by interacting alginic acid with sodium carbonate in alcohol, then adding the calcium salt, drying the mixture and grinding the residue to a fine powder. The further addition of a buffering agent, such as tetrasodium pyrophosphate, and more calcium as calcium gluconate gave a smooth pudding gel within 15 min. Of course, the use of milk in this mixture, while leading no doubt to a better pudding, upsets the nutritional nullity of the mixture.

Alginates have also been successfully employed as ice-cream stabilisers, ingredients of processed cheese, whipped cream, as coatings for frozen fish or chicken, as synthetic sausage skins, icings and toppings for cakes, as 'chiffon pie' fillings and ingredients of salad dressings.[8] A dramatic omen of things to come for the British, addicted as they are to "chips with everything", are the patents[9] for synthetic potato chips. These are made from a wet dispersion of sodium alginate, starch, potato flour and calcium lactate. This is

[7] Gibson, K. F., U.S. Patent, 2,808,337.
[8] Glecksman, M. (1962), *Adv. Food Research*, 11, 138.
[9] Rivoche, E., U.S. Patent 2,786,763; 2,786,764; 2,791,508.

extruded continuously through a nozzle and chopped off into chip-sized lengths which are then fried in the customary manner in a chip-pan. A further example of what can be done by the suitable application of chemical understanding to make artificial non-nutritious foods is provided by the successful development of synthetic cherries. These synthetic fruit, devised as long ago as 1946,[10] are made by allowing drops of a mixture of suitably coloured and flavoured sodium alginate solution to fall into a bath containing a solution of an appropriate calcium salt; calcium chloride is a popular choice. A skin of insoluble calcium alginate forms at once on the outside of each drop. If the drops are then allowed to 'cure' for a period of time, the calcium ions gradually penetrate into the centre and cause the whole thing to gel. The advantage claimed for the artificial cherries is that they are not affected by the heat of the baking oven when they are used to decorate cakes.

Cherries in fruit cakes, whether they are artificial or not, may not appear to be of much importance when viewed in the light of human hunger, or perhaps of human obesity either. Nevertheless, they are a portent of what can be done. Such artificial 'fruit' are indeed widely used in fruit cake and pie fillings. They have been successfully marketed in Holland, France, Italy, Switzerland, Finland—and in Australia.[11] A similar product prepared in the United States is made by machine in five sizes, from 'buckshot' up to three-quarters of an inch in diameter.

It can be argued that where deficiency disease like beri-beri kill their hundreds, obesity, due to a superfluity of food, shortens the lives of thousands. Be that as it may, 'slimming' is big business. Intelligent fat men—and fat girls as well—know that they eat too much. The problem, which the books on diet, the 'health' foods and the systems of exercises and regimen all attempt to solve, is how can the patients be steeled to refuse food. Perhaps the solution lies in the chemical synthesis of non-foods from which attractive meals of guaranteed physiological valuelessness can be constructed.

One final application of the alginate molecule in combination with cellulose, also without nutritional value, but made eatable by chemical modification into methylethylcellulose, is the manufacture of syn-

[10] Peschardt, W. J. S., U.S. Patent, 2,403,547.
[11] Anon. (1961), *Food Process.*, 22, 50.

thetic cream.[12] This can be made by a combination containing only about 0·15% of sodium alginate and 0·5% of methylethylcellulose. This mixture is claimed to possess very satisfactory whipping properties. But however satisfactory such a product may be as filling for an eclair or as topping for a trifle, it is clearly useless for frying. True fats and oils are, of all food components, the richest in calorific value. Since, as was pointed out in Chapter 4, hydrocarbons are not absorbed to any extent, it was at one time suggested that liquid paraffin might be used as a readily available non-nutritional substitute for cooking fat or frying oil. Indeed, at one time it was used as an ingredient of 'non-fattening' mayonnaise and other like commodities. Edible fats are available to the body as food because the long-chain chemical structures of the fatty acid molecules of which they are principally composed are linked to the glycerol unit to which they are combined in a way which can be broken by the digestive enzymes and subsequently metabolised by the body. Paraffin chains, on the other hand, possess no such point of attack for the digestive or metabolic enzymes. It was the general acceptance of their non-availability that led to the use of liquid paraffin, first as a digestive lubricant and then as a non-absorbable food ingredient. It was, however, subsequently discovered[13] that paraffins, when they are finely emulsified, as is done when they are used to make mayonnaise, are in fact absorbed, at least to a small extent. It appears that absorbed mineral oil, although it is not metabolised, may accumulate in the cells of the liver and spleen[14] and may even land up in the lungs.[15] Mineral paraffins are, therefore, not generally used as artificial fat, although in Great Britain[16] up to 5% may be added to dried fruit and they can be employed as a coating on eggs and on the rind of cheese. In the United States it was calculated that in 1966 more than 47 g of paraffin were being consumed per head of the population[14] annually.

The undesirable effects due to the absorption of traces of hydrocarbon oils, however small, and their subsequent accumulation in

[12] Anon. (1958), *Food Manuf.* 33, 159.

[13] El Mahdi, M. A. H. (1933), and Channon, H. J., *Biochem. J.*, 27, 1487; Channon, H. J. (1933), and Devine, J., *Biochem. J.*, 28, 467; Frazer, A. C. (1942), and Stewart, H. C., *Nature*, 149, 167.

[14] Boitnott, J. K. (1966), and Margolis, S., *Bull Johns Hopkins Hosp.*, 118, 402, 414.

[15] Eisenberg, R. F. (1966), and Oatway, W. H., *Calif. Med.*, 105, 214.

[16] *Mineral Hydrocarbons in Food Regulations* (1966), (Stat. Inst. 1073).

the liver and elsewhere, must undoubtedly be taken seriously. Nevertheless, it would appear to be a research target fully capable of attainment to discover a mineral oil of such molecular size and configuration as not to be capable of absorption yet which would nevertheless be suitable for use as a non-nutritional frying medium for non-foods.

If we reflect on the whole topic of the synthesis of non-nutritional food-stuffs we can recognise that it is not merely frivolous, as might first appear, but illustrates two important principles of nutritional science. The first is the question of the biochemical principles which underlie molecular structures essential for foods to be utilisable by the body. A fatty acid chain terminating thus:

$$\begin{array}{ccccccccc} & H & & H & & H & & O & \\ & | & & | & & | & & \| & \\ - & C & - & C & - & C & - & C & -OH \\ & | & & | & & | & & & \\ & H & & H & & H & & & \end{array}$$

can be absorbed by the body and utilised as biological fuel. A hydrocarbon chain of mineral oil ending thus:

$$\begin{array}{ccccccccc} & H & & H & & H & & H & \\ & | & & | & & | & & | & \\ - & C & - & C & - & C & - & C & -H \\ & | & & | & & | & & | & \\ & H & & H & & H & & H & \end{array}$$

is not absorbed (or only in trace—if troublesome—amounts) and cannot be used as biological fuel. Or again, a polysaccharide chain of starch is so linked as to be readily available as food. The polysaccharide chain of cellulose, differently linked, but made up of the same glucose units as starch, is not metabolisable by the human system.

I should perhaps add in parentheses that certain micro-organisms do possess an enzyme, cellulase, which enables them to metabolise cellulose. Recently, pharmaceutical manufacturers in Japan have succeeded in preparing this enzyme on a commercial scale. It is now fashionable for Japanese gentlemen, at the end of a large meal, to take a tablet of cellulase to aid the digestion of any cellulose which, in the form of stringy cabbage or paper from the bottom of a cake, they may accidentally have eaten. The possibilities opened up by the Japanese achievement in preparing cellulase on a large scale from microbiological cultures are quite remarkable to the speculative mind. It could, for example, render fibrous material, or even wood itself, edible. While the possibility of a man being able to eat straw—or even his straw hat—is not likely, such material could, however, become a source of food.

The second serious lesson to be learned from the chemistry of non-foods is that nutrition, while it is primarily concerned with the chemical composition of the metabolisable nutrients diet contributes to the body's needs, is equally concerned with the dietary behaviour of the people who select and eat certain commodities and refuse others. Colour and flavour, like consistency, lightness, crispness and warmth, contribute nothing to the chemical adequacy of the diet, just as disgust at the idea of eating horse, dog flesh or mice detract nothing from the nutritional excellence of these items. Yet the behaviour of the men and women who make up a community possessing a particular culture is relevant to any serious study of nutrition and food supply. The culture of an increasing proportion of the world's population is that of industrialisation based on science and technology. Within such a culture, the synthesis of non-foods may well play a significant part.

CHAPTER 10

FOOD MANUFACTURE

The studies which I have reviewed in this book show quite clearly that there is already ample scientific information to enable food to be manufactured by chemical synthesis from inedible raw materials. Already each one of a long list of vitamins is synthesised on an industrial scale for addition to staple foods sold at prices within the reach of ordinary people. By this means, the nutritional needs of the consumers are met at an economic cost—and this is the point. It is a rational procedure for public-health authorities to decree that synthetic thiamine should be incorporated in white bread rather than lay down a law to insist that only brown bread should be sold. The laws of economics insist that what people want they are prepared to pay for; and the laws of economics, which deal with people's desires, are equally the concern of the student of human behaviour as are the laws of nutrition, which deal with the biochemical demands of the individual cells of which human beings are composed.

After having reviewed the falling costs of synthesising such nutrients as thiamine, vitamin A, ascorbic acid and of the amino-acids, lysine, threonine and methionine, A. T. McPherson[1] calculated the order of cost at which food synthesis was being achieved. While accepting that at the time he was writing (1960) not all food components could be produced by chemical synthesis, he prophesied—and the evidence of the successive chapters of this book suggests that his prophecy is likely to be fulfilled—that in due course this would become possible. He then proceeded to estimate that in order to synthesise enough food to supply the entire needs of the 53 million people, by which number the world's population is at present increasing each year, an investment of $13 000 million in chemical plant would be needed annually. While this is undoubtedly a lot of chemical plant, for which year by year substantial areas of land, services, waste-disposal facilities and trained manpower would be required, it is only seven

[1] McPherson, A. T. (1960), *J. Wash. Acad. Sci.*, 50, 1; — (1961), *Ind. Res.*, Nov., 21.

times the current annual investment of the US chemical industry, and is small in comparison with the sums spent on armaments.

It is true that chemical science enables the chemical industry to achieve remarkable results. What the industrial chemist actually does, however, is a combination of what he is capable of doing by chemistry and what he is enabled or encouraged to do—I must repeat—by economics, which economics, the gloomy science, is an attribute of biology in as much as it reflects the behaviour and the desires of the human tribe. It deals, in fact, with an essential feature of the natural history of man.

In 1914, the population of Great Britain spent, on average, 60% of their money income on food. In 1956, 22 years later, they spent only 40% of their money on food. By 1967, the proportion of their expenditure devoted to food had fallen to 30%. Thus, in the scale of values, the importance of food had been halved in 53 years. And within the category of 'food' itself, expenditure on separate items had changed equally strikingly. The proportion of the total food expenditure devoted to cereals, the cheapest source of calories, fell, while the proportion spent on meat and on more sophisticated items increased. And, to study the matter in more detail still, we find in recent years the relative importance of dehydrated and, particularly, of frozen foods, becoming greater. Whereas in Scotland, to take one example, the cost of oatmeal at the farm was, in 1969, say 1p per lb and ground oats little more, these commodities enjoyed little popularity. Even prepared 'porridge' oats at three or four times the price were in dwindling demand. Such oatmeal as was eaten tended to be 'instant' porridge costing some fifteen times as much as the oatmeal itself.

This increase in retail price is a simple indication of the fact, sometimes overlooked by those whose interest is in food science and the nutritional well-being of particular communities, that food must be judged as an economic commodity like anything else. Even under the most primitive conditions for people living at a level legitimately described as 'bare subsistence', the necessity for food must always be balanced against demands for shelter, clothing or, for that matter, cigarettes and a seat at the cinema if the community should be in the West, or for, say, a bicycle, if the group under consideration happened to be in Central Africa. But, of course, the majority of the world's population enjoy an economic status—and it is economic status that for the most part determines the state of nutrition—above mere subsistence. There are, furthermore, two worlds, and in the technologically developed world the economic level is very high indeed.

The annual review of the Food and Agriculture Organisation of the United Nations for 1968,[2] in contrast to earlier and more gloomy reports, gave figures to show that, whereas during the decade 1955–65 the annual increase in the world's population was about 2·5%, and the yearly increase in food production about the same, in 1967, the increase in agricultural production in the developing countries was 6%. The significance of this remarkable rise in food production lies in the fact that in that year, for the first time, a significant proportion of crops was derived from improved seeds of high-yielding strains of cereals.[3] These strains had been developed at the International Maize and Wheat Improvement Centre in Mexico and at the International Rice Research Centre in the Philippines. Whereas standard varieties of wheat were yielding 1·2–3·0 tons per hectare, the new dwarf varieties, Sonora 64 and Lerma Pojo 64, yield from 4 to 8 tons. Equally large increases in yield were obtained following the introduction of the new, improved 1R8 strain of rice. In 1966–67, about 500 000 hectares of land in Pakistan and 800 000 hectares in India were sown with the new Senora 64 and Lerma Pojo 64 strains of wheat. Turkey used 22 400 tons of the improved seed and, among other countries, Afghanistan, Iran, Iraq, Kenya, Lebanon, Libya, Morocco, Rhodesia, South Africa and Tunisia all made a start in growing it.

Application of the new rice had an equally dramatic effect on increasing the yield of food from the soil. It was widely sown in the Philippines, where it was first developed, and in India as well. As a result, the amount of rice produced in the great rice-growing district of Tanjore was increased by an estimated 450 000 tons.

Besides wheat and rice, there are also new strains of maize, sorghum and potatoes from which there is good reason to hope that equally big improvements in yield can be expected in Latin America. And apart from the remarkable improvement in the production of food which has actually been obtained in 1968, there is scope for even more important advances. For example, whereas the average yields of wheat in Great Britain and the Netherlands are 40·5 and 44·4 quintals per hectare, that in India is 8·3, and in Iraq only 4·6. Even

[2] F.A.O. (1968), *The State of Food and Agriculture 1968*. Rome.
[3] Anon. (1969), *Nature, Lond.*, 223, 118.

in Australia and the United States they only amount to 12·0 and 17·4 quintals per hectare. It can be seen, therefore, that not only are striking increases in food production and in the adequacy of food supplies compared with the increasing populations of the world being obtained, but further major improvements are possible even assuming no further development occurs in better seeds.

The future significance of synthetic food must always depend on the adequacy of the supplies of natural food. Whether synthesised foods make a significant contribution will also be affected by their relative price compared with natural food. This presents a somewhat paradoxical situation. If natural foods are cheap, it may well not prove to be economic to manufacture synthetic ones. But when world prices for foods are low, producers obtain little encouragement to improve their methods and there is a temptation for some farmers to restrict their activities to subsistence farming sufficient only to feed their own families. Perhaps, therefore, synthetic foods may, like margarine, come into their own when the high prices for natural foods tempt growers to increase the amounts they produce, but which at the same time leads consumers to welcome supplies of something cheaper. Several factors may, however, affect this simple system. For example, artificial foods, such as margarine and simulated meat fabricated from spun vegetable protein, are designed to be as similar as possible in taste, appearance and consistency to butter and real meat. It is curious to note that artificial flowers made from plastic, which could be designed in novel and exotic shapes unknown to Nature, are also made to resemble daffodils, tulips and other common natural blooms. Indeed, so accurate are the reproductions that it is often difficult to distinguish between natural and artificial petals without touching them. Although the preparation of food items made primarily from synthetic ingredients is so far in an early stage of development, there is no reason why artificial foods should not be different from natural foods and might, it is hoped, be accepted as better and more delightful.

It is sometimes argued that the whole idea of synthesising food is misconceived and that it is technologically prodigal to manufacture food artificially when we could obtain it without drawing on our credit of banked energy by making use, instead, of the rays of the sun. It is, in essence, this energy—the energy of sunlight—which is the nourishment we obtain from the bread we eat. The sun and the part of its energy which gives us life are, from the point of view of human existence, everlasting and come to us afresh every morning. The carbon compounds used as basic raw materials for chemically synthesised foods are, on the other hand, derived from exhaustible stocks of coal or, as is more likely, petroleum. Our views on the exhaustibility of petroleum have, it is true, radically changed in the last two generations. As one oil-field after another is discovered and exploited it becomes clear that, although the supplies of oil must be finite, they are undoubtedly very large. And the argument that to make use of mined material is foolish because it will eventually all be used up is a bad argument. It could, for example, be used equally to argue in favour of wooden beams in buildings rather than steel girders. As a footnote to this argument, it is interesting to note that the abundance of petroleum is so great and the chemical versatility of the hydrocarbons of which it is composed is so remarkable—for the manufacture of all sorts of polymers, whether they be plastics, as generally understood, or synthetic food—that it has been forecast[4] that by the mid-1980s the volume of plastics produced will exceed that of steel.

Schemes for producing yeast grown on petroleum as a source of protein, compared with protein prepared from leaf juice, run into this problem of whether or not petroleum can justifiably be used as a source of food. One way of answering the question is to compare the costs of the two. When this is done, it must be accepted that the cost of the petroleum fraction needed for yeast growth is less, when the basic cost per ton after purification, transport and storage have been taken into account, than the cost of leaves. The leaves themselves, discarded pea haulms, perhaps, beet tops or water hyacinth recovered from an artificial lake or waterway, may cost little or nothing. The cost of collection to run a protein-extraction plant, combined with the difficulty or impossibility of holding a stock in storage, substantially inflate the expense of the operation so that the cost per ton of protein from 'free' leaves may be greater than that of a ton of protein from purchased petroleum.

A further argument that is also sometimes raised in discussing the feasibility of synthesising food is that the energy input needed to make the food ought not to exceed the amount of food energy eventually obtained. This argument seems to me to be invalid also.

[4] Iliff, N. (1968), *Chem. & Ind.*, 1064.

In 1947, J. C. D. Hutchinson[5] carried out a laborious and carefully planned scientific study to discover the efficiency with which rabbit meat could be produced from weeds. Litters of weaned rabbits were brought into the experiment at intervals from the last week of May until the middle of July in order to cover the season when weeds are plentiful in England. In due course the carcasses of the rabbits were analysed and the increase in the edible matter determined in terms of protein and fat, and from these figures the yield of calories was calculated. The average yield of gross energy value per weed-fed rabbit was 549 calories. Hutchinson, being a trained physiologist and a devoted scientist to boot, also carried out a study of the energy which he himself expended in walking about with a sack collecting 15 lb of weeds to feed to the rabbits. This amounted to 1022 calories. Curious and interesting though the results of this experiment may be, the apparently obvious conclusion that to raise rabbits in this way makes a negative contribution to human diet is an over simplification. Hutchinson himself remarked that the value of the operation depended on the relative value to man of available calories and of animal protein. But values in human society are far more complex and subtle even than this. The Cubans, after having considered their agricultural policy afresh since their revolution, have concluded that cigar tobacco is for them a crop better fitted to increase the prosperity, and hence the nutritional status, of their population than dairying.

The artificial synthesis of food is an example of the substitution of chemical manufacture for what had previously been a biological process. This replacement of biology by chemistry has already been achieved for rubber. The small amount of expensive 'methyl rubber' made in Germany in 1918 was followed by the American invention of 'Thiokol', derived from petroleum and sulphur, in 1922. By 1940, the massive efforts of the du Pont and Nemours Company had achieved the synthesis of about 2000 tons of so-called Neoprene, while in Germany the 115 000 tons of Buna S, made from butadiene and styrene produced from coal, was good enough for the manufacture of tyres.[6] Today, the great chemical plants built both in Germany and in the United States under the stress of war continue to produce at an economic price synthetic rubber from what must be acknowledged to be eventually exhaustable stocks of fossil fuel

[5] Hutchinson, J. C. D. (1947), *Brit. J. Nutr.*, 1, 231.

[6] Bauer, P. T. (1948), *The Rubber Industry*. Longmans, London.

and use it mingled with supplies of natural rubber from presumably everlasting rubber trees.

The story of artificial—so-called 'man-made'—fibres is the same. Again, we find biological products—cotton and flax from self-perpetuating crops, wool from continuously reproducing sheep, and silk from everlasting worms—being replaced by chemical products increasingly derived from petroleum. Synthetic fibres for clothing are now produced in competition—that is, in constantly self-adjusting economic equilibrium—with fibres derived from biological sources. de Chardonnet began his study of 'artificial silk' in 1878. After a hundred years of scientific research, technological effort and active marketing, man-made fibres accounted for 54% of the consumption of all apparel fibres in the United States.[7]

The time-scale for synthetic food is likely to be quicker once the necessity of full-scale research and development is recognised by governments and the giant commercial interests. We have seen that the chemical synthesis of the basic components of food can already be achieved; that substantial amounts of vitamins and, more recently, of amino-acids are being synthesised and marketed on a commercial basis, as well as flavours, colours, emulsifiers and various sorts of 'improvers'.

What will be next—the use of petroleum for the synthesis of fat or even sugar, or the synthesis of complete protein? The possibilities for the future are immense.

[7] Anon. (1968), *Chem. & Eng.*, 1538.

INDEX